Telecommunications Essentials

Telecommunications Essentials

Hudson Warner

NY RESEARCH PRESS
New York

Published by NY Research Press
118-35 Queens Blvd., Suite 400,
Forest Hills, NY 11375, USA
www.nyresearchpress.com

Telecommunications Essentials
Hudson Warner

International Standard Book Number: 978-1-63238-865-0 (Hardback)

Cataloging-in-Publication Data

Telecommunications essentials / Hudson Warner.
p. cm.
Includes bibliographical references and index.
ISBN 978-1-63238-865-0
1. Telecommunication. 2. Communication. 3. Telecommuting. I. Warner, Hudson.
TK5101 .T45 2022
621.382--dc23

Table of Contents

Preface

This book has been written, keeping in view that students want more practical information. Thus, my aim has been to make it as comprehensive as possible for the readers. I would like to extend my thanks to my family and co-workers for their knowledge, support and encouragement all along.

The transmission of signals, signs, messages, images, words and sounds by wire, radio, optical and other electromagnetic systems is known as telecommunications. The information is transmitted via transmission media or through electromagnetic radiation. The transmission paths are divided into communication channels in order to facilitate multiplexing. The technologies of telecommunication are broadly divided into wireless and wired methods. Some of the common media which make use of principles from telecommunications are telephone, radio, television, internet, local area network, etc. The three main elements of a telecommunication system are transmitter, transmission medium and receiver area. Transmitter is used to take information and convert it to a signal. The transmission medium carries the signal, and the receiver takes the signal from the channel and converts it to usable information for the recipient. The topics included in this book on telecommunications are of utmost significance and bound to provide incredible insights to readers. It covers in detail some existing theories and innovative concepts related to this field. This textbook attempts to assist those with a goal of delving into this field.

A brief description of the chapters is provided below for further understanding:

Chapter - Introduction

The exchange of information by electronic and electrical means over significant distances is referred to as telecommunications. A few of the common telecommunications devices are telephones, telegraph, radio, fiber optics, satellites, etc. This chapter has been carefully written to provide an easy understanding of the diverse aspects of telecommunications.

Chapter - Transmission Media and Technologies

There are various mediums through which media and technologies are transmitted. Some of these mediums are wire transmissions, radio waves, coxial cables, fiber optic communication, etc. This chapter will briefly introduce all the significant aspects of these transmission media and technologies.

Chapter - Network Topology and Switching

Network Topology can be defined as the layout of the network which connects different nodes of a network to establish a connection. The hardware device that channels incoming data from multiple input ports to a specific output port that will take it towards its final destination is known as a network switch. All the varied aspects of network topology and switching have been carefully analyzed in this chapter.

Chapter - Multiplexing and Multiple Access Techniques

Multiplexing is a method by which multiple digital signals are combined into one signal over a shared medium. The techniques which allow multiple mobile users to share the allotted spectrum in the most efficient manner are known as multiple access techniques. The topics elaborated in this chapter will help in gaining a better perspective about multiplexing and multiple access techniques.

Chapter - Telecommunications Networks

The collection of terminal nodes which are linked to enable telecommunication between the terminals is referred to as a telecommunication network. Some of the common telecommunication networks include television network, cellular network, computer network, etc. This chapter has been carefully written to provide an easy understanding of these telecommunication networks.

Hudson Warner

1

Introduction

The exchange of information by electronic and electrical means over significant distances is referred to as telecommunications. A few of the common telecommunications devices are telephones, telegraph, radio, fiber optics, satellites, etc. This chapter has been carefully written to provide an easy understanding of the diverse aspects of telecommunications.

TELECOMMUNICATIONS

Telecommunication is the science and practice of transmitting information by electromagnetic means. Modern telecommunication centres on the problems involved in transmitting large volumes of information over long distances without damaging loss due to noise and interference. The basic components of a modern digital telecommunications system must be capable of transmitting voice, data, radio, and television signals. Digital transmission is employed in order to achieve high reliability and because the cost of digital switching systems is much lower than the cost of analog systems. In order to use digital transmission, however, the analog signals that make up most voice, radio, and television communication must be subjected to a process of analog-to-digital conversion. (In data transmission this step is bypassed because the signals are already in digital form; most television, radio, and voice communication, however, use the analog system and must be digitized.) In many cases, the digitized signal is passed through a source encoder, which employs a number of formulas to reduce redundant binary information. After source encoding, the digitized signal is processed in a channel encoder, which introduces redundant information that allows errors to be detected and corrected. The encoded signal is made suitable for transmission by modulation onto a carrier wave and may be made part of a larger signal in a process known as multiplexing. The multiplexed signal is then sent into a multiple-access transmission channel. After transmission, the above process is reversed at the receiving end, and the information is extracted.

Analog-to-Digital Conversion

In transmission of speech, audio, or video information, the object is high fidelity—that is, the best possible reproduction of the original message without the degradations imposed by signal distortion and noise. The basis of relatively noise-free and distortion-free telecommunication is the binary signal. The simplest possible signal of any kind that can be employed to transmit messages, the binary signal consists of only two possible values. These values are represented by the binary digits, or bits, 1 and 0. Unless the noise and distortion picked up during transmission are great enough to change the binary signal from one value to another, the correct value can be determined by the receiver so that perfect reception can occur.

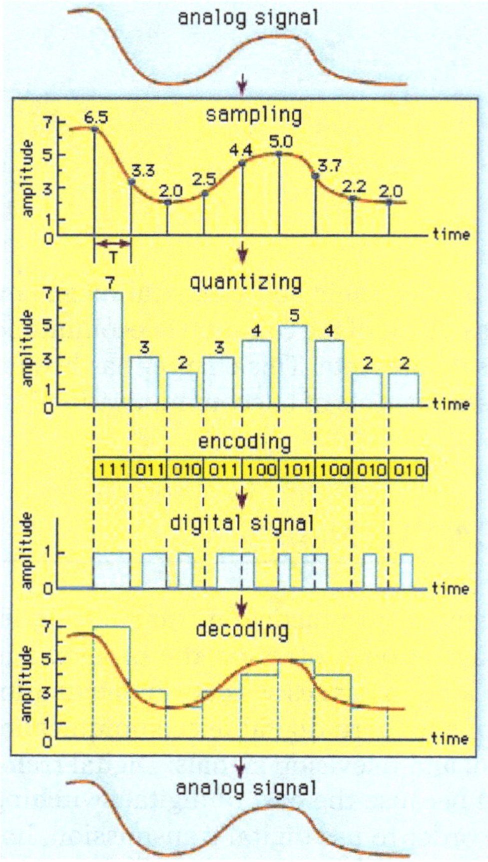

Basic steps in analog-to-digital conversion.

An analog signal is sampled at regular intervals. The amplitude at each interval is quantized, or assigned a value, and the values are mapped into a series of binary digits, or bits. The information is transmitted as a digital signal to the receiver, where it is decoded and the analog signal reconstituted.

If the information to be transmitted is already in binary form (as in data communication), there is no need for the signal to be digitally encoded. But ordinary voice communications taking place by way of a telephone are not in binary form; neither is much of the information gathered for transmission from a space probe, nor are the television or radio signals gathered for transmission through a satellite link. Such signals, which continually vary among a range of values, are said to be analog, and in digital communications systems analog signals must be converted to digital form. The process of making this signal conversion is called analog-to-digital (A/D) conversion.

Sampling

Analog-to-digital conversion begins with sampling, or measuring the amplitude of the analog waveform at equally spaced discrete instants of time. The fact that samples of a continually varying wave may be used to represent that wave relies on the assumption that the wave is constrained in its rate of variation. Because a communications signal is actually a complex wave—essentially the sum of a number of component sine waves, all of which have their own precise amplitudes and phases—the rate of variation of the complex wave can be measured by the frequencies of oscillation of all its components. The difference between the maximum rate of oscillation (or highest frequency) and the minimum rate of oscillation (or lowest frequency) of the sine waves making up the signal is known as the bandwidth (B) of the signal. Bandwidth thus represents the maximum frequency range occupied by a signal. In the case of a voice signal having a minimum frequency of 300 hertz and a maximum frequency of 3,300 hertz, the bandwidth is 3,000 hertz, or 3 kilohertz. Audio signals generally occupy about 20 kilohertz of bandwidth, and standard video signals occupy approximately 6 million hertz, or 6 megahertz.

The concept of bandwidth is central to all telecommunication. In analog-to-digital conversion, there is a fundamental theorem that the analog signal may be uniquely represented by discrete samples spaced no more than one over twice the bandwidth (1/2B) apart. This theorem is commonly referred to as the sampling theorem, and the sampling interval (1/2B seconds) is referred to as the Nyquist interval (after the Swedish-born American electrical engineer Harry Nyquist). As an example of the Nyquist interval, in past telephone practice the bandwidth, commonly fixed at 3,000 hertz, was sampled at least every 1/6,000 second. In current practice 8,000 samples are taken per second, in order to increase the frequency range and the fidelity of the speech representation.

Quantization

In order for a sampled signal to be stored or transmitted in digital form, each sampled amplitude must be converted to one of a finite number of possible values, or levels. For ease in conversion to binary form, the number of levels is usually a power of 2—that is, 8, 16, 32, 64, 128, 256, and so on, depending on the degree of precision required. In digital transmission of voice, 256 levels are commonly used because tests have shown that this provides adequate fidelity for the average telephone listener.

The input to the quantizer is a sequence of sampled amplitudes for which there are an infinite number of possible values. The output of the quantizer, on the other hand, must be restricted to a finite number of levels. Assigning infinitely variable amplitudes to a limited number of levels inevitably introduces inaccuracy, and inaccuracy results in a corresponding amount of signal distortion. (For this reason quantization is often called a "lossy" system.) The degree of inaccuracy depends on the number of output levels used by the quantizer. More quantization levels increase the accuracy of the representation, but they also increase the storage capacity or transmission speed required. Better performance with the same number of output levels can be achieved by judicious placement of the output levels and the amplitude thresholds needed for assigning those levels. This placement in turn depends on the nature of the waveform that is being quantized. Generally, an optimal quantizer places more levels in amplitude ranges where the signal is more likely to occur and fewer levels where the signal is less likely. This technique is known as nonlinear quantization. Nonlinear quantization can also be accomplished by passing the signal through a compressor circuit, which amplifies the signal's weak components and attenuates its strong components. The compressed signal, now occupying a narrower dynamic range, can be quantized with a uniform, or linear, spacing of thresholds and output levels. In the case of the telephone signal, the compressed signal is uniformly quantized at 256 levels, each level being represented by a sequence of eight bits.

At the receiving end, the reconstituted signal is expanded to its original range of amplitudes. This sequence of compression and expansion, known as companding, can yield an effective dynamic range equivalent to 13 bits.

Bit Mapping

In the next step in the digitization process, the output of the quantizer is mapped into a binary sequence. An encoding table that might be used to generate the binary sequence is shown below:

quantization level	binary code
0	000
1	001
2	010
3	011
4	100
5	101
6	110
7	111

It is apparent that 8 levels require three binary digits, or bits; 16 levels require four bits; and 256 levels require eight bits. In general 2^n levels require n bits.

In the case of 256-level voice quantization, where each level is represented by a sequence of 8 bits, the overall rate of transmission is 8,000 samples per second times 8 bits per sample, or 64,000 bits per second. All 8 bits must be transmitted before the next sample appears. In order to use more levels, more binary samples would have to be squeezed into the allotted time slot between successive signal samples. The circuitry would become more costly, and the bandwidth of the system would become correspondingly greater. Some transmission channels (telephone wires are one example) may not have the bandwidth capability required for the increased number of binary samples and would distort the digital signals. Thus, although the accuracy required determines the number of quantization levels used, the resultant binary sequence must still be transmitted within the bandwidth tolerance allowed.

Source Encoding

As is pointed out in analog-to-digital conversion, any available telecommunications medium has a limited capacity for data transmission. This capacity is commonly measured by the parameter called bandwidth. Since the bandwidth of a signal increases with the number of bits to be transmitted each second, an important function of a digital communications system is to represent the digitized signal by as few bits as possible—that is, to reduce redundancy. Redundancy reduction is accomplished by a source encoder, which often operates in conjunction with the analog-to-digital converter.

Huffman Codes

In general, fewer bits on the average will be needed if the source encoder takes into account the probabilities at which different quantization levels are likely to occur. A simple example will illustrate this concept. Assume a quantizing scale of only four levels: 1, 2, 3, and 4. Following the usual standard of binary encoding, each of the four levels would be mapped by a two-bit code word. But also assume that level 1 occurs 50 percent of the time, that level 2 occurs 25 percent of the time, and that levels 3 and 4 each occur 12.5 percent of the time. Using variable-bit code words might cause more efficient mapping of these levels to be achieved. The variable-bit encoding rule would use only one bit 50 percent of the time, two bits 25 percent of the time, and three bits 25 percent of the time. On average it would use 1.75 bits per sample rather than the 2 bits per sample used in the standard code.

Thus, for any given set of levels and associated probabilities, there is an optimal encoding rule that minimizes the number of bits needed to represent the source. This encoding rule is known as the Huffman code, after the American D.A. Huffman, who created it in 1952. Even more efficient encoding is possible by grouping sequences of levels together and applying the Huffman code to these sequences.

The Lempel-ziv Algorithm

The design and performance of the Huffman code depends on the designers' knowing the probabilities of different levels and sequences of levels. In many cases, however, it is desirable to have an encoding system that can adapt to the unknown probabilities of a source. A very efficient technique for encoding sources without needing to know their probable occurrence was developed in the 1970s by the Israelis Abraham Lempel and Jacob Ziv. The Lempel-Ziv algorithm works by constructing a codebook out of sequences encountered previously. For example, the codebook might begin with a set of four 12-bit code words representing four possible signal levels. If two of those levels arrived in sequence, the encoder, rather than transmitting two full code words (of length 24), would transmit the code word for the first level (12 bits) and then an extra two bits to indicate the second level. The encoder would then construct a new code word of 12 bits for the sequence of two levels, so that even fewer bits would be used thereafter to represent that particular combination of levels. The encoder would continue to read quantization levels until another sequence arrived for which there was no code word. In this case the sequence without the last level would be in the codebook, but not the whole sequence of levels. Again, the encoder would transmit the code word for the initial sequence of levels and then an extra two bits for the last level. The process would continue until all 4,096 possible 12-bit combinations had been assigned as code words.

In practice, standard algorithms for compressing binary files use code words of 12 bits and transmit 1 extra bit to indicate a new sequence. Using such a code, the Lempel-Ziv algorithm can compress transmissions of English text by about 55 percent, whereas the Huffman code compresses the transmission by only 43 percent.

Run-length Codes

Certain signal sources are known to produce "runs," or long sequences of only 1s or 0s. In these cases it is more efficient to transmit a code for the length of the run rather than all the bits that represent the run itself. One source of long runs is the fax machine. A fax machine works by scanning a document and mapping very small areas of the document into either a black pixel (picture element) or a white pixel. The document is divided into a number of lines (approximately 100 per inch), with 1,728 pixels in each line (at standard resolution). If all black pixels were mapped into 1s and all white pixels into 0s, then the scanned document would be represented by 1,857,600 bits (for a standard American 11-inch page). At older modem transmission speeds of 4,800 bits per second, it would take 6 minutes 27 seconds to send a single page. If, however, the sequence of 0s and 1s were compressed using a run-length code, significant reductions in transmission time would be made.

The code for fax machines is actually a combination of a run-length code and a Huffman code; it can be explained as follows: A run-length code maps run lengths into code words, and the codebook is partitioned into two parts. The first part contains symbols for runs of lengths that are a multiple of 64; the second part is made up of runs from 0 to 63 pixels. Any run length would then be represented as a multiple of 64 plus some remainder. For example, a run of 205 pixels would be sent using the code word for a run of length 192 (3 × 64) plus the code word for a run of length 13. In this way the number of bits needed to represent the run is decreased significantly. In addition, certain runs that are known to have a higher probability of occurrence are encoded into code words of short length, further reducing the number of bits that need to be transmitted. Using this type of encoding, typical compressions for facsimile transmission range between 4 to 1 and 8 to 1. Coupled to higher modem speeds, these compressions reduce the transmission time of a single page to between 48 seconds and 1 minute 37 seconds.

Channel Encoding

As described in Source encoding, one purpose of the source encoder is to eliminate redundant binary digits from the digitized signal. The strategy of the channel encoder, on the other hand, is to add redundancy to the transmitted signal—in this case so that errors caused by noise during transmission can be corrected at the receiver. The process of encoding for protection against channel errors is called error-control coding. Error-control codes are used in a variety of applications, including satellite communication, deep-space communication, mobile radio communication, and computer networking.

There are two commonly employed methods for protecting electronically transmitted information from errors. One method is called forward error control (FEC). In this method information bits are protected against errors by the transmitting of extra redundant bits, so that if errors occur during transmission the redundant bits can be used by the decoder to determine where the errors have occurred and how to correct them. The second method of error control is called automatic repeat request (ARQ). In this method redundant bits are added to the transmitted information and are used by the receiver to detect errors. The receiver then signals a request for a repeat transmission. Generally, the number of extra bits needed simply to detect an error, as in the ARQ system, is much smaller than the number of redundant bits needed both to detect and to correct an error, as in the FEC system.

Repetition Codes

One simple, but not usually implemented, FEC method is to send each data bit three times. The receiver examines the three transmissions and decides by majority vote whether a 0 or 1 represents a sample of the original signal. In this coded system, called a repetition code of block-length three and rate one-third, three times as many bits per second are used to transmit the same signal as are used by an uncoded system; hence, for a fixed available bandwidth only one-third as many signals can be conveyed with the coded system as compared with the uncoded system. The gain is that now at least two of the three coded bits must be in error before a reception error occurs.

The Hamming Code

Another simple example of an FEC code is known as the Hamming code. This code is able to protect a four-bit information signal from a single error on the channel by adding three redundant bits to the signal. Each sequence of seven bits (four information bits plus three redundant bits) is called a code word. The first redundant bit is chosen so that the sum of ones in the first three information bits plus the first redundant bit amounts to an even number. (This calculation is called a parity check, and the redundant bit is called a parity bit.) The second parity bit is chosen so that the sum of the ones in the last three information bits plus the second parity bit is even, and the third parity bit is chosen so that the sum of ones in the first, second, and fourth information bits and the last parity bit is even. This code can correct a single channel error by recomputing the parity checks. A parity check that fails indicates an error in one of the positions checked, and the two subsequent parity checks, by process of elimination, determine the precise location of the error. The Hamming code thus can correct any single error that occurs in any of the seven positions. If a double error occurs, however, the decoder will choose the wrong code word.

Convolutional Encoding

The Hamming code is called a block code because information is blocked into bit sequences of finite length to which a number of redundant bits are added. When k information bits are provided to a block encoder, n − k redundancy bits are appended to the information bits to form a transmitted code word of n bits. The entire code word of length n is thus completely determined by one block of k information bits. In another channel-encoding scheme, known as convolutional encoding, the encoder output is not naturally segmented into blocks but is instead an unending stream of bits. In convolutional encoding, memory is incorporated into the encoding process, so that the preceding M blocks of k information bits, together with the current block of k information bits, determine the encoder output. The encoder accomplishes this by shifting among a finite number of "states," or "nodes." There are several variations of convolutional encoding, but the simplest example may be seen in what is known as the (n,1) encoder, in which the current block of k information bits consists of only one bit. At each given state of the (n,1) encoder, when the information bit (a 0 or a 1) is received, the encoder transmits a sequence of n bits assigned to represent that bit when the encoder is at that current state. At the same time, the encoder shifts to one of only two possible successor states, depending on whether the information bit was a 0 or a 1. At this successor state, in turn, the next information bit is represented by a specific sequence of n bits, and the encoder is again shifted to one of two possible successor states. In this way, the sequence of information bits stored in the encoder's memory determines both the state of the encoder and its output, which is modulated and transmitted across the channel. At the receiver, the demodulated bit sequence is compared to the possible bit sequences that can be produced by the encoder. The receiver determines the bit sequence that is most likely to have been transmitted, often by using an efficient decoding algorithm called Viterbi decoding (after its inventor, A.J. Viterbi). In general, the greater the memory (i.e., the more states) used by the encoder, the better the error-correcting performance of the code—but only at the cost of a more complex decoding algorithm. In addition, the larger the number of bits (n) used to transmit information, the better the performance—at the cost of a decreased data rate or larger bandwidth.

Coding and decoding processes similar to those described above are employed in trellis coding, a coding scheme used in high-speed modems. However, instead of the sequence of bits that is produced by a convolutional encoder, a trellis encoder produces a sequence of modulation symbols. At the transmitter, the channel-encoding process is coupled with the modulation process, producing a system known as trellis-coded modulation. At the receiver, decoding and demodulating are performed jointly in order to optimize the performance of the error-correcting algorithm.

Modulation

In many telecommunications systems, it is necessary to represent an information-bearing signal with a waveform that can pass accurately through a transmission medium. This assigning of a suitable waveform is accomplished by modulation, which is the process by which some characteristic of a carrier wave is varied in accordance with an information signal, or modulating wave. The modulated signal is then transmitted over a channel, after which the original information-bearing signal is recovered through a process of demodulation.

Modulation is applied to information signals for a number of reasons, some of which are outlined below.

1. Many transmission channels are characterized by limited passbands—that is, they will pass only certain ranges of frequencies without seriously attenuating them (reducing their amplitude). Modulation methods must therefore be applied to the information signals in order to "frequency translate" the signals into the range of frequencies that are permitted by the channel. Examples of channels that exhibit passband characteristics include alternating-current-coupled coaxial cables, which pass signals only in the range of 60 kilohertz to several hundred megahertz, and fibre-optic cables, which pass light signals only within a given wavelength range without significant attenuation. In these instances frequency translation is used to "fit" the information signal to the communications channel.
2. In many instances a communications channel is shared by multiple users. In order to prevent mutual interference, each user's information signal is modulated onto an assigned carrier of a specific frequency. When the frequency assignment and subsequent combining is done at a central point, the resulting combination is a frequency-division multiplexed signal, as is discussed in Multiplexing. Frequently there is no central combining point, and the communications channel itself acts as a distributed combine. An example of the latter situation is the broadcast radio bands (from 540 kilohertz to 600 megahertz), which permit simultaneous transmission of multiple AM radio, FM radio, and television signals without mutual interference as long as each signal is assigned to a different frequency band.
3. Even when the communications channel can support direct transmission of the information-bearing signal, there are often practical reasons why this is undesirable. A simple example is the transmission of a three-kilohertz (i.e., voiceband) signal via radio wave. In free space the wavelength of a three-kilohertz signal is 100 kilometres (60 miles). Since an effective radio antenna is typically as large as half the wavelength of the signal, a three-kilohertz radio wave might require an antenna up to 50 kilometres in length. In this case translation of the voice frequency to a higher frequency would allow the use of a much smaller antenna.

Analog Modulation

As is noted in analog-to-digital conversion, voice signals, as well as audio and video signals, are inherently analog in form. In most modern systems these signals are digitized prior to transmission, but in some systems the analog signals are still transmitted directly without converting them to digital form. There are two commonly used methods of modulating analog signals. One technique, called amplitude modulation, varies the amplitude of a fixed-frequency carrier wave in proportion to the information signal. The other technique, called frequency modulation, varies the frequency of a fixed-amplitude carrier wave in proportion to the information signal.

Digital Modulation

In order to transmit computer data and other digitized information over a communications channel, an analog carrier wave can be modulated to reflect the binary nature of the digital baseband signal. The parameters of the carrier that can be modified are the amplitude, the frequency, and the phase.

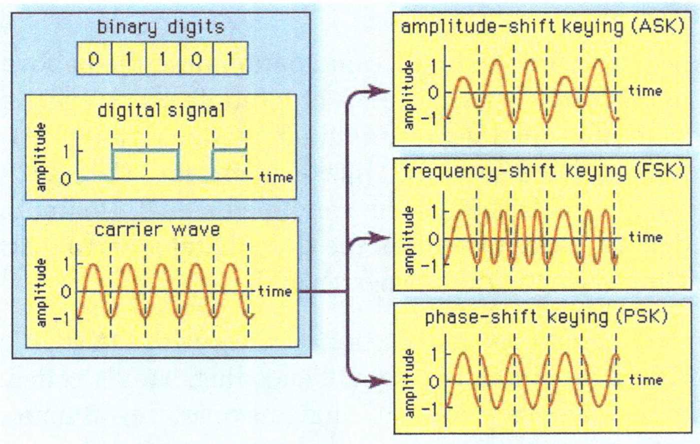

Three methods of digital signal modulation.

A digital signal, representing the binary digits 0 and 1 by a series of on and off amplitudes, is impressed onto an analog carrier wave of constant amplitude and frequency. In amplitude-shift keying (ASK), the modulated wave represents the series of bits by shifting abruptly between high and low amplitude. In frequency-shift keying (FSK), the bit stream is represented by shifts between two frequencies. In phase-shift keying (PSK), amplitude and frequency remain constant; the bit stream is represented by shifts in the phase of the modulated signal.

Amplitude-shift Keying

If amplitude is the only parameter of the carrier wave to be altered by the information signal, the modulating method is called amplitude-shift keying (ASK). ASK can be considered a digital version of analog amplitude modulation. In its simplest form, a burst of radio frequency is transmitted only when a binary 1 appears and is stopped when a 0 appears. In another variation, the 0 and 1 are represented in the modulated signal by a shift between two preselected amplitudes.

Frequency-shift Keying

If frequency is the parameter chosen to be a function of the information signal, the modulation method is called frequency-shift keying (FSK). In the simplest form of FSK signaling, digital data is transmitted using one of two frequencies, whereby one frequency is used to transmit a 1 and the other frequency to transmit a 0. Such a scheme was used in the Bell 103 voiceband modem, introduced in 1962, to transmit information at rates up to 300 bits per second over the public switched telephone network. In the Bell 103 modem, frequencies of 1,080 +/- 100 hertz and 1,750 +/- 100 hertz were used to send binary data in both directions.

Phase-shift Keying

When phase is the parameter altered by the information signal, the method is called phase-shift keying (PSK). In the simplest form of PSK a single radio frequency carrier is sent with a fixed phase to represent a 0 and with a 180° phase shift—that is, with the opposite polarity—to represent a 1. PSK was employed in the Bell 212 modem, which was introduced about 1980 to transmit information at rates up to 1,200 bits per second over the public switched telephone network.

Advanced Methods

In addition to the elementary forms of digital modulation described above, there exist more advanced methods that result from a superposition of multiple modulating signals. An example of the latter form of modulation is quadrature amplitude modulation (QAM). QAM signals actually transmit two amplitude-modulated signals in phase quadrature (i.e., 90° apart), so that four or more bits are represented by each shift of the combined signal. Communications systems that employ QAM include digital cellular systems in the United States and Japan as well as most voiceband modems transmitting above 2,400 bits per second.

A form of modulation that combines convolutional codes with QAM is known as trellis-coded modulation (TCM), which is described in Channel encoding. Trellis-coded modulation forms an essential part of most of the modern voiceband modems operating at data rates of 9,600 bits per second and above, including V.32 and V.34 modems.

2

Transmission Media and Technologies

There are various mediums through which media and technologies are transmitted. Some of these mediums are wire transmissions, radio waves, coxial cables, fiber optic communication, etc. This chapter will briefly introduce all the significant aspects of these transmission media and technologies.

TELECOMMUNICATIONS MEDIA

Telecommunications media are the equipment and systems (metal wire, terrestrial and satellite radio, and optical fibre) employed in the transmission of electromagnetic signals.

Radio wave dish-type antennas, varying in diameter from 8 to 30 metres (26 to 98 feet), serving an Earth station in a satellite communications network.

Transmission Media and the Problem of Signal Degradation

Every telecommunications system involves the transmission of an information-bearing electromagnetic signal through a physical medium that separates the transmitter from the receiver. All transmitted signals are to some extent degraded by the environment through which they propagate. Signal degradation can take many forms, but generally it falls into three types: noise, distortion, and attenuation (reduction in power). Noise is the presence of random, unpredictable, and undesirable electromagnetic emissions that can mask the intended information signal. Distortion is any undesired change in the amplitude or phase of any component of an information signal that causes a change in the overall waveform of the signal. Both noise and distortion are commonly introduced by all transmission media, and they both result in errors in reception. The relative impact of these factors on reliable communication depends on the rate of information transmission, on the desired fidelity upon reception, and on whether communication must occur in "real time"—i.e., as in telephone conversations and video teleconferencing.

Various modulating and encoding schemes have been devised to provide protection against the errors caused by channel distortion and channel noise. In addition to these signal-processing techniques, protection against reception errors can be provided by boosting the power of the transmitter, thus increasing the signal-to-noise ratio (the ratio of signal power to noise power). However, even powerful signals suffer some degree of attenuation as they pass through the transmission medium. The principal cause of power loss is dissipation, the conversion of part of the electromagnetic energy to another form of energy such as heat. In communications media, channel attenuation is typically expressed in decibels (dB) per unit distance. Attenuation of zero decibels means that the signal is passed without loss; three decibels means that the power of the signal decreases by one-half. The plot of channel attenuation as the signal frequency is varied is known as the attenuation spectrum, while the average attenuation over the entire frequency range of a transmitted signal is defined as the attenuation coefficient.

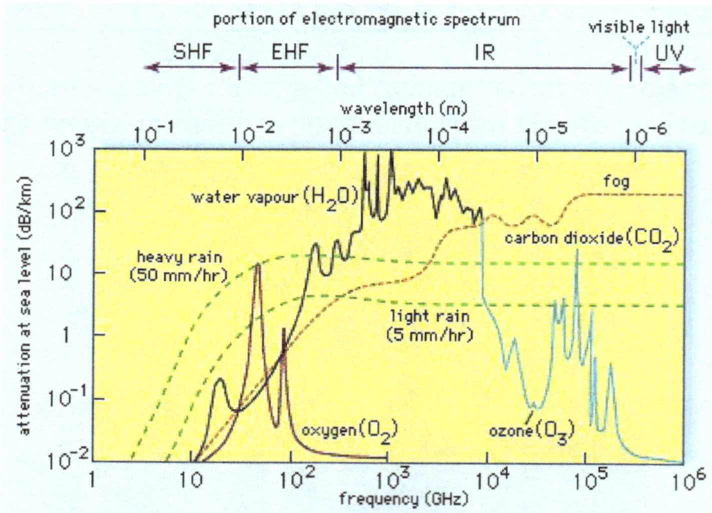

Attenuation of electromagnetic energy propagated through the atmosphere at sea level along a horizontal path. A broad range of the attenuation spectrum is shown, from microwave radiowaves to ultraviolet light.

Channel attenuation is an important factor in the use of each transmission medium. Along with noise and distortion, it can influence the choice of one medium over another.

TRANSMITTER

In electronics and telecommunications a transmitter or radio transmitter is an electronic device which produces radio waves with an antenna. The transmitter itself generates a radio frequency alternating current, which is applied to the antenna. When excited by this alternating current, the antenna radiates radio waves.

Transmitters are necessary component parts of all electronic devices that communicate by radio, such as radio and television broadcasting stations, cell phones, walkie-talkies, wireless computer networks, Bluetooth enabled devices, garage door openers, two-way radios in aircraft, ships, spacecraft, radar sets and navigational beacons. The term transmitter is usually limited to equipment that generates radio waves for communication purposes; or radiolocation, such as radar and navigational transmitters. Generators of radio waves for heating or industrial purposes, such as microwave ovens or diathermy equipment, are not usually called transmitters, even though they often have similar circuits.

The term is popularly used more specifically to refer to a broadcast transmitter, a transmitter used in broadcasting, as in FM radio transmitter or television transmitter. This usage typically includes both the transmitter proper, the antenna, and often the building it is housed in.

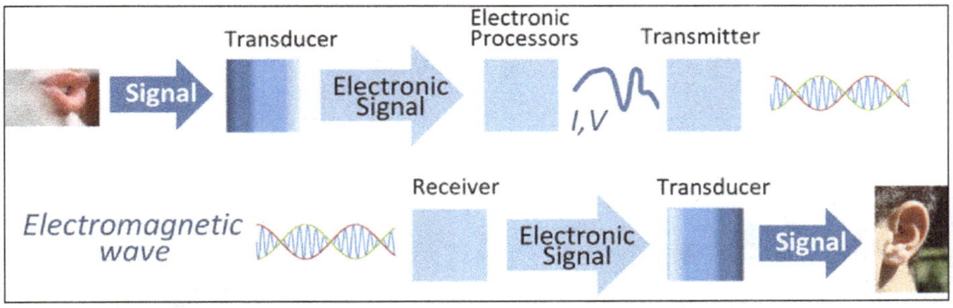

A radio transmitter is usually part of a radio communication system which uses electromagnetic waves (radio waves) to transport information (in this case sound) over a distance.

A transmitter can be a separate piece of electronic equipment, or an electrical circuit within another electronic device. A transmitter and a receiver combined in one unit is called a transceiver. The term transmitter is often abbreviated "XMTR" or "TX" in technical documents. The purpose of most transmitters is radio communication of information over a distance. The information is provided to the transmitter in the form of an electronic signal, such as an audio (sound) signal from a microphone, a video (TV) signal from a video camera, or in wireless networking devices, a digital signal from a computer. The transmitter combines the information signal to be carried with the radio frequency signal which generates the radio waves, which is called the carrier signal. This process is called *modulation*. The information can be added to the carrier in several different ways, in different types of transmitters. In an amplitude modulation (AM) transmitter, the information is added to the radio signal by varying its amplitude. In a frequency modulation (FM) transmitter, it is added by varying the radio signal's frequency slightly. Many other types of modulation are also used.

35 kW, Continental 816R-5B FM transmitter, belonging to American FM radio station KWNR broadcasting on 95.5 MHz in Las Vegas.

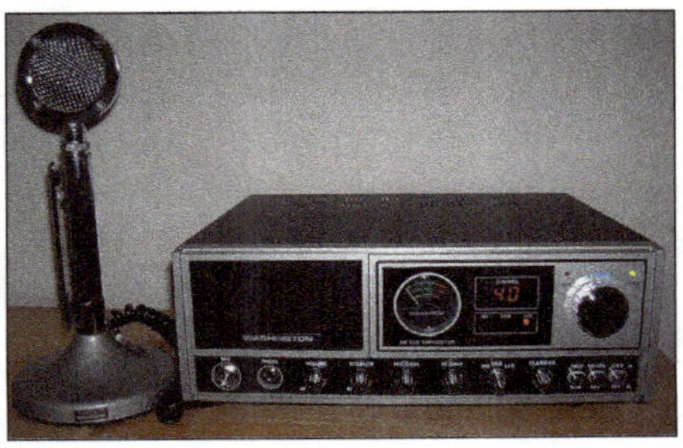

A CB radio transceiver, a two way radio transmitting on 27 MHz with a power of 4 W, that can be operated without a license.

Modern amateur radio transceiver, the ICOM IC-746PRO. It can transmit on the amateur bands from 1.8 MHz to 144 MHz with an output power of 100 W.

The radio signal from the transmitter is applied to the antenna, which radiates the energy as radio waves. The antenna may be enclosed inside the case or attached to the outside of the transmitter, as in portable devices such as cell phones, walkie-talkies, and garage door openers. In more powerful transmitters, the antenna may be located on top of a building or on a separate tower, and connected to the transmitter by a feed line, that is a transmission line.

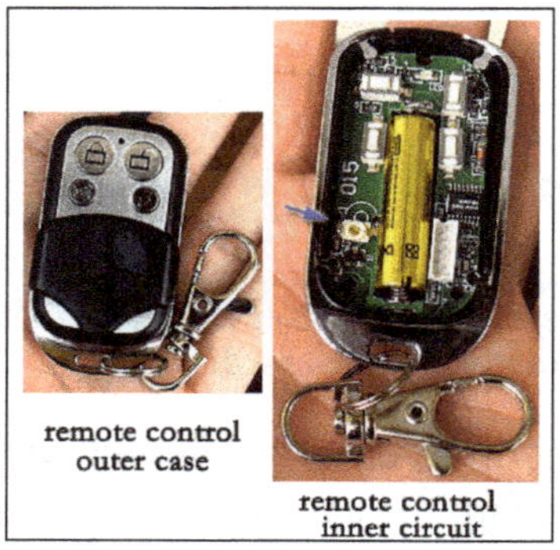

A garage door opener control contains a low-power 2.4 GHz transmitter that sends coded commands to the garage door mechanism to open or close.

A cellphone has several transmitters: a duplex cell transceiver, a Wi-Fi modem, and a Bluetooth modem.

In a wireless computer network, wireless routers like this contain a 2.4 GHz transmitter that sends and receives network packets for computers on the local area network.

Operation

The antenna in the center is two vertical metal rods, with an alternating current applied at its center from a radio transmitter *(not shown)*. The voltage charges the two sides of the antenna alternately positive *(+)* and negative *(–)*. Loops of electric field *(black lines)* leave the antenna and travel away at the speed of light; these are the radio waves.

Electromagnetic waves are radiated by electric charges when they are accelerated. Radio waves, electromagnetic waves of radio frequency, are generated by time-varying electric currents, consisting of electrons flowing through a metal conductor called an antenna which are changing their velocity or direction and thus accelerating. An alternating current flowing back and forth in an antenna will create an oscillating magnetic field around the conductor. The alternating voltage will also charge the ends of the conductor alternately positive and negative, creating an oscillating electric field around the conductor. If the frequency of the oscillations is high enough, in the radio frequency range above about 20 kHz, the oscillating coupled electric and magnetic fields will radiate away from the antenna into space as an electromagnetic wave, a radio wave.

A radio transmitter is an electronic circuit which transforms electric power from a power source into a radio frequency alternating current to apply to the antenna, and the antenna radiates the energy from this current as radio waves. The transmitter also impresses information such as an audio or video signal onto the radio frequency current to be carried by the radio waves. When they strike the antenna of a radio receiver, the waves excite similar (but less powerful) radio frequency currents in it. The radio receiver extracts the information from the received waves.

Components

A practical radio transmitter usually consists of these parts:

- In high power transmitters, a power supply circuit to transform the input electrical power to the higher voltages needed to produce the required power output.
- An electronic oscillator circuit to generate the radio frequency signal. This usually generates a sine wave of constant amplitude called the carrier wave, because it serves to "carry" the information through space. In most modern transmitters, this is a crystal oscillator in which the frequency is precisely controlled by the vibrations of a quartz crystal. The frequency of the carrier wave is considered the frequency of the transmitter.
- A modulator circuit to add the information to be transmitted to the carrier wave produced by the oscillator. This is done by varying some aspect of the carrier wave. The information is provided to the transmitter either in the form of an audio signal, which represents sound, a video signal which represents moving images, or for data in the form of a binary digital signal which represents a sequence of bits, a bitstream. Different types of transmitters use different modulation methods to transmit information:
    - In an AM (amplitude modulation) transmitter the amplitude (strength) of the carrier wave is varied in proportion to the modulation signal.
    - In an FM (frequency modulation) transmitter the frequency of the carrier is varied by the modulation signal.
    - In an FSK (frequency-shift keying) transmitter, which transmits digital data, the frequency of the carrier is shifted between two frequencies which represent the two binary digits, 0 and 1.

- OFDM (orthogonal frequency division multiplexing) is a family of complicated digital modulation methods very widely used in high bandwidth systems such as WiFi networks, cellphones, digital television broadcasting, and digital audio broadcasting (DAB) to transmit digital data using a minimum of radio spectrum bandwidth. OFDM has higher spectral efficiency and more resistance to fading than AM or FM. In OFDM multiple radio carrier waves closely spaced in frequency are transmitted within the radio channel, with each carrier modulated with bits from the incoming bitstream so multiple bits are being sent simultaneously, in parallel. At the receiver the carriers are demodulated and the bits are combined in the proper order into one bitstream.

Many other types of modulation are also used. In large transmitters the oscillator and modulator together are often referred to as the *exciter*.

- A radio frequency (RF) amplifier to increase the power of the signal, to increase the range of the radio waves.
- An impedance matching (antenna tuner) circuit to match the impedance of the transmitter to the impedance of the antenna (or the transmission line to the antenna), to transfer power efficiently to the antenna. If these impedances are not equal, it causes a condition called standing waves, in which the power is reflected back from the antenna toward the transmitter, wasting power and sometimes overheating the transmitter.

In higher frequency transmitters, in the UHF and microwave range, free running oscillators are unstable at the output frequency. Older designs used an oscillator at a lower frequency, which was multiplied by frequency multipliers to get a signal at the desired frequency. Modern designs more commonly use an oscillator at the operating frequency which is stabilized by phase locking to a very stable lower frequency reference, usually a crystal oscillator.

Legal Restrictions

In most parts of the world, use of transmitters is strictly controlled by law because of the potential for dangerous interference with other radio transmissions (for example to emergency communications). Transmitters must be licensed by governments, under a variety of license classes depending on use such as broadcast, marine radio, Airband, Amateur and are restricted to certain frequencies and power levels. A body called the International Telecommunications Union (ITU) allocates the frequency bands in the radio spectrum to various classes of users. In some classes, each transmitter is given a unique call sign consisting of a string of letters and numbers which must be used as an identifier in transmissions. The operator of the transmitter usually must hold a government license, such as a general radiotelephone operator license, which is obtained by passing a test demonstrating adequate technical and legal knowledge of safe radio operation.

Exceptions to the above regulations allow the unlicensed use of low-power short-range transmitters in consumer products such as cell phones, cordless telephones, wireless microphones, walkie-talkies, Wi-Fi and Bluetooth devices, garage door openers, and baby monitors. In the US, these fall under Part 15 of the Federal Communications Commission (FCC) regulations. Although they can be operated without a license, these devices still generally must be type-approved before sale.

RADIO RECEIVER

In radio communications, a radio receiver, also known as a receiver, wireless or simply radio is an electronic device that receives radio waves and converts the information carried by them to a usable form. It is used with an antenna. The antenna intercepts radio waves (electromagnetic waves) and converts them to tiny alternating currents which are applied to the receiver, and the receiver extracts the desired information. The receiver uses electronic filters to separate the desired radio frequency signal from all the other signals picked up by the antenna, an electronic amplifier to increase the power of the signal for further processing, and finally recovers the desired information through demodulation.

The information produced by the receiver may be in the form of sound, moving images (television), or data. A radio receiver may be a separate piece of electronic equipment, or an electronic circuit within another device. Radio receivers are very widely used in modern technology, as components of communications, broadcasting, remote control, and wireless networking systems. In consumer electronics, the terms *radio* and *radio receiver* are often used specifically for receivers designed to reproduce sound transmitted by radio broadcasting stations, historically the first mass-market commercial radio application.

Broadcast Radio Receivers

The most familiar form of radio receiver is a broadcast receiver, often just called a *radio*, which receives audio programs intended for public reception transmitted by local radio stations. The sound is reproduced either by a loudspeaker in the radio or an earphone which plugs into a jack on the radio. The radio requires electric power, provided either by batteries inside the radio or a power cord which plugs into an electric outlet. All radios have a volume control to adjust the loudness of the audio, and some type of "tuning" control to select the radio station to be received.

Modulation Types

Modulation is the process of adding information to a radio carrier wave.

AM and FM

Two types of modulation are used in analog radio broadcasting systems; AM and FM.

In amplitude modulation (AM) the strength of the radio signal is varied by the audio signal. AM broadcasting is allowed in the AM broadcast bands which are between 148 and 283 kHz in the longwave range, and between 526 and 1706 kHz in the medium frequency (MF) range of the radio spectrum. AM broadcasting is also permitted in shortwave bands, between about 2.3 and 26 MHz, which are used for long distance international broadcasting.

In frequency modulation (FM) the frequency of the radio signal is varied slightly by the audio signal. FM broadcasting is permitted in the FM broadcast bands between about 65 and 108 MHz in the very high frequency (VHF) range. The exact frequency ranges vary somewhat in different countries.

FM stereo radio stations broadcast in stereophonic sound (stereo), transmitting two sound channels representing left and right microphones. A stereo receiver contains the additional circuits and parallel signal paths to reproduce the two separate channels. A monaural receiver, in contrast, only receives a single audio channel that is a combination (sum) of the left and right channels. While AM stereo transmitters and receivers exist, they have not achieved the popularity of FM stereo.

Most modern radios are "AM/FM" radios, and are able to receive both AM and FM radio stations, and have a switch to select which band to receive.

Digital Audio Broadcasting

Digital audio broadcasting (DAB) is an advanced radio technology which debuted in some countries in 1998 that transmits audio from terrestrial radio stations as a digital signal rather than an analog signal as AM and FM do. Its advantages are that DAB has the potential to provide higher quality sound than FM (although many stations do not choose to transmit at such high quality), has greater immunity to radio noise and interference, makes better use of scarce radio spectrum bandwidth, and provides advanced user features such as electronic program guide, sports commentaries, and image slideshows. Its disadvantage is that it is incompatible with previous radios so that a new DAB receiver must be purchased. As of 2017, 38 countries offer DAB, with 2,100 stations serving listening areas containing 420 million people. Most countries plan an eventual switchover from FM to DAB. The United States and Canada have chosen not to implement DAB.

DAB radio stations work differently from AM or FM stations: a single DAB station transmits a wide 1,500 kHz bandwidth signal that carries from 9 to 12 channels from which the listener can choose. Broadcasters can transmit a channel at a range of different bit rates, so different channels can have different audio quality. In different countries DAB stations broadcast in either Band III (174–240 MHz) or L band (1.452–1.492 GHz).

Reception

The signal strength of radio waves decreases the farther they travel from the transmitter, so a radio station can only be received within a limited range of its transmitter. The range depends on the power of the transmitter, the sensitivity of the receiver, atmospheric and internal noise, as well as any geographical obstructions such as hills between transmitter and receiver. AM broadcast band radio waves travel as ground waves which follow the contour of the Earth, so AM radio stations can be reliably received at hundreds of miles distance. Due to their higher frequency, FM band radio signals cannot travel far beyond the visual horizon; limiting reception distance to about 40 miles (64 km), and can be blocked by hills between the transmitter and receiver. However FM radio is less susceptible to interference from radio noise (RFI, sferics, static) and has higher fidelity; better frequency response and less audio distortion, than AM. So in many countries serious music is only broadcast by FM stations, and AM stations specialize in radio news, talk radio, and sports. Like FM, DAB signals travel by line of sight so reception distances are limited by the visual horizon to about 30–40 miles (48–64 km).

Types of Broadcast Receiver

Radios are made in a range of styles and functions:

- Table radio: A self-contained radio with speaker designed to sit on a table.
- Clock radio: A bedside table radio that also includes an alarm clock. The alarm clock can be set to turn on the radio in the morning instead of an alarm, to wake the owner.
- Tuner: A high fidelity AM/FM radio receiver in a component home audio system. It has no speakers but outputs an audio signal which is fed into the system and played through the system's speakers.
- Portable radio: A radio powered by batteries that can be carried with a person. Radios are now often integrated with other audio sources in CD players and portable media players.
    - Boom box: A portable battery-powered high fidelity stereo sound system in the form of a box with a handle, which became popular during the mid 1970s.
    - Transistor radio: An older term for a portable pocket-sized broadcast radio receiver. Made possible by the invention of the transistor and developed in the 1950s, transistor radios were hugely popular during the 1960s and early 1970s, and changed the public's listening habits.
- Car radio: An AM/FM radio integrated into the dashboard of a vehicle, used for entertainment while driving. Virtually all modern cars and trucks are equipped with radios, which usually also includes a CD player.
- Satellite radio receiver: Subscription radio receiver that receives audio programming from a direct broadcast satellite. The subscriber must pay a monthly fee. They are mostly designed as car radios.
- Shortwave receiver: This is a broadcast radio that also receives the shortwave bands. It is used for shortwave listening.
- AV receivers are a common component in a high-fidelity or home-theatre system; in addition to receiving radio programming, the receiver will also contain switching and amplifying functions to interconnect and control the other components of the system.

Other Applications

Radio receivers are essential components of all systems that use radio. Besides broadcast receivers, described above, radio receivers are used in a huge variety of electronic systems in modern technology. They can be a separate piece of equipment (a *radio*), or a subsystem incorporated into other electronic devices. A transceiver is a transmitter and receiver combined in one unit. Below is a list of a few of the most common types, organized by function.

- Broadcast television reception: Televisions receive a video signal representing a moving image, composed of a sequence of still images, and a synchronized audio signal representing the associated sound. The television channel received by a TV occupies a wider bandwidth than an audio signal, from 600 kHz to 6 MHz.

- Terrestrial television receiver, broadcast television or just television (TV): Televisions contains an integral receiver (TV tuner) which receives free broadcast television from local television stations on TV channels in the VHF and UHF bands.
- Satellite TV receiver: A set-top box which receives subscription direct-broadcast satellite television, and displays it on an ordinary television. A rooftop satellite dish receives many channels all modulated on a K_u band microwave downlink signal from a geostationary direct broadcast satellite 22,000 miles (35,000 km) above the Earth, and the signal is converted to a lower intermediate frequency and transported to the box through a coaxial cable. The subscriber pays a monthly fee.

- Two-way voice communications: A two-way radio is an audio transceiver, a receiver and transmitter in the same device, used for bidirectional person-to-person voice communication. The radio link may be half-duplex, using a single radio channel in which only one radio can transmit at a time. so different users take turns talking, pressing a push to talk button on their radio which switches on the transmitter. Or the radio link may be full duplex, a bidirectional link using two radio channels so both people can talk at the same time, as in a cell phone.
    - Cellphone: A portable telephone that is connected to the telephone network by radio signals exchanged with a local antenna called a cell tower. Cellphones have highly automated digital receivers working in the UHF and microwave band that receive the incoming side of the duplex voice channel, as well as a control channel that handles dialing calls and switching the phone between cell towers. They usually also have several other receivers that connect them with other networks: a WiFi modem, a bluetooth modem, and a GPS receiver. The cell tower has sophisticated multichannel receivers that receive the signals from many cell phones simultaneously.
    - Cordless phone: A landline telephone in which the handset is portable and communicates with the rest of the phone by a short range duplex radio link, instead of being attached by a cord. Both the handset and the base station have radio receivers operating in the UHF band that receive the short range bidirectional duplex radio link.
    - Citizens band radio: A two-way half-duplex radio operating in the 27 MHz band that can be used without a license. They are often installed in vehicles and used by truckers and delivery services.
    - Walkie-talkie: A handheld short range half-duplex two-way radio.

Handheld scanner.

- Scanner: A receiver that continuously monitors multiple frequencies or radio channels by stepping through the channels repeatedly, listening briefly to each channel for a transmission. When a transmitter is found the receiver stops at that channel. Scanners are used to monitor emergency police, fire, and ambulance frequencies, as well as other two way radio frequencies such as citizens band. Scanning capabilities have also become a standard feature in communications receivers, walkie-talkies, and other two-way radios.

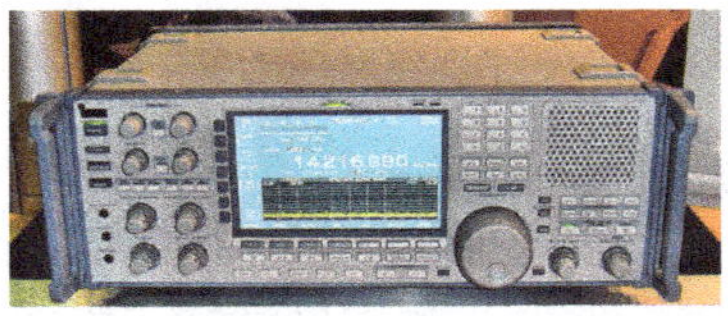

Modern communications receiver, ICOM RC-9500.

- Communications receiver or shortwave receiver: A general purpose audio receiver covering the LF, MF, shortwave (HF), and VHF bands. Used mostly with a separate shortwave transmitter for two-way voice communication in communication stations, amateur radio stations, and for shortwave listening.

- One-way (simplex) voice communications:
  - Wireless microphone receiver: These receive the short range signal from wireless microphones used onstage by musical artists, public speakers, and television personalities.

Baby monitor. The receiver is on the left.

  - Baby monitor: This is a cribside appliance for mothers of infants that transmits the baby's sounds to a receiver carried by the mother, so she can monitor the baby while she is in other parts of the house. Many baby monitors now have video cameras to show a picture of the baby.

- Data communications:
  - Wireless (WiFi) modem: An automated short range digital data transmitter and receiver on a portable wireless device that communicates by microwaves with a nearby access point, a router or gateway, connecting the portable device with a local computer network (WLAN) to exchange data with other devices.
  - Bluetooth modem: A very short range (up to 10 m) 2.4-2.83 GHz data transceiver on a portable wireless device used as a substitute for a wire or cable connection, mainly to

exchange files between portable devices and connect cellphones and music players with wireless earphones.

- Microwave relay: A long distance high bandwidth point-to-point data transmission link consisting of a dish antenna and transmitter that transmits a beam of microwaves to another dish antenna and receiver. Since the antennas must be in line-of-sight, distances are limited by the visual horizon to 30–40 miles. Microwave links are used for private business data, wide area computer networks (WANs), and by telephone companies to transmit distance phone calls and television signals between cities.

- Satellite communications: Communication satellites are used for data transmission between widely separated points on Earth. Other satellites are used for search and rescue, remote sensing, weather reporting and scientific research. Radio communication with satellites and spacecraft can involve very long path lengths, from 35,786 km (22,236 mi) for geosynchronous satellites to billions of kilometers for interplanetary spacecraft. This and the limited power available to a spacecraft transmitter mean very sensitive receivers must be used.
    - Satellite transponder: A receiver and transmitter in a communications satellite that receives multiple data channels carrying long distance telephone calls, television signals. or internet traffic on a microwave uplink signal from a satellite ground station and retransmits the data to another ground station on a different downlink frequency. In a direct broadcast satellite the transponder broadcasts a stronger signal directly to satellite radio or satellite television receivers in consumer's homes.
    - Satellite ground station receiver: Communication satellite ground stations receive data from communications satellites orbiting the Earth. Deep space ground stations such as those of the NASA Deep Space Network receive the weak signals from distant scientific spacecraft on interplanetary exploration missions. These have large dish antennas around 85 ft (25 m) in diameter, and extremely sensitive radio receivers similar to radio telescopes. The RF front end of the receiver is often cryogenically cooled to −195.79 °C (−320 °F) by liquid nitrogen to reduce radio noise in the circuit.
- Remote control: Remote control receivers receive digital commands that control a device, which may be as complex as a space vehicle or unmanned aerial vehicle, or as simple as a garage door opener. Remote control systems often also incorporate a telemetry channel to transmit data on the state of the controlled device back to the controller. Radio controlled model and other models include multichannel receivers in model cars, boats, airplanes, and helicopters. A short-range radio system is used in keyless entry systems.
- Radiolocation: This is the use of radio waves to determine the location or direction of an object.
    - Radar: A device that transmits a narrow beam of microwaves which reflect from a target back to a receiver, used to locate objects such as aircraft, spacecraft, missiles, ships or land vehicles. The reflected waves from the target are received by a receiver usually connected to the same antenna, indicating the direction to the target. Widely used in aviation, shipping, navigation, weather forecasting, space flight, vehicle collision avoidance systems, and the military.

- Global navigation satellite system (GNSS) receiver, such as a GPS receiver used with the US Global Positioning System: The most widely used electronic navigation device. An automated digital receiver that receives simultaneous data signals from several satellites in low Earth orbit. Using extremely precise time signals it calculates the distance to the satellites, and from this the receiver's location on Earth. GNSS receivers are sold as portable devices, and are also incorporated in cell phones, vehicles and weapons, even artillery shells.
- VOR receiver: Navigational instrument on an aircraft that uses the VHF signal from VOR navigational beacons between 108 and 117.95 MHz to determine the direction to the beacon very accurately, for air navigation.
- Wild animal tracking receiver: a receiver with a directional antenna used to track wild animals which have been tagged with a small VHF transmitter, for wildlife management purposes.

- Other
    - Telemetry receiver: This receives data signals to monitor conditions of a process. Telemetry is used to monitor missile and spacecraft in flight, well logging during oil and gas drilling, and unmanned scientific instruments in remote locations.
    - Measuring receiver: A calibrated, laboratory grade radio receiver used to measure the characteristics of radio signals. Often incorporates a spectrum analyzer.
    - Radio telescope: specialized antenna and radio receiver used as a scientific instrument to study weak radio waves from astronomical radio sources in space like stars, nebulas and galaxies in radio astronomy. They are the most sensitive radio receivers that exist, having large parabolic (dish) antennas up to 500 meters in diameter, and extremely sensitive radio circuits. The RF front end of the receiver is often cryogenically cooled by liquid nitrogen to reduce radio noise.

Working of Receiver

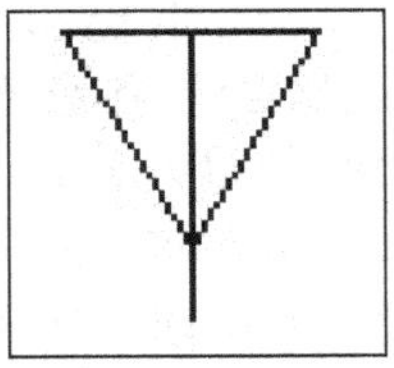

Symbol for an antenna.

A radio receiver is connected to an antenna which converts some of the energy from the incoming radio wave into a tiny radio frequency AC voltage which is applied to the receiver's input. An antenna typically consists of an arrangement of metal conductors. The oscillating electric and magnetic fields of the radio wave push the electrons in the antenna back and forth, creating an oscillating voltage.

The antenna may be enclosed inside the receiver's case, as with the ferrite loop antennas of AM radios and the flat inverted F antenna of cell phones; attached to the outside of the receiver, as with whip antennas used on FM radios, or mounted separately and connected to the receiver by a cable, as with rooftop television antennas and satellite dishes.

Filtering, Amplification and Demodulation

Practical radio receivers perform three basic functions on the signal from the antenna: filtering, amplification, and demodulation:

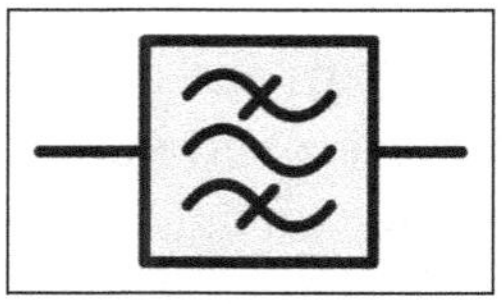

Symbol for a bandpass filter used in block diagrams of radio receivers.

- **Bandpass filtering:** Radio waves from many transmitters pass through the air simultaneously without interfering with each other. These can be separated in the receiver because they have different frequencies; that is, the radio wave from each transmitter oscillates at a different rate. To separate out the desired radio signal, the bandpass filter allows the frequency of the desired radio transmission to pass through, and blocks signals at all other frequencies.

  The bandpass filter consists of one or more resonant circuits (tuned circuits). The resonant circuit is connected between the antenna input and ground. When the incoming radio signal is at the resonant frequency, the resonant circuit has high impedance and the radio signal from the desired station is passed on to the following stages of the receiver. At all other frequencies the resonant circuit has low impedance, so signals at these frequencies are conducted to ground.

  - **Bandwidth and selectivity:** The information (modulation) in a radio transmission is contained in two narrow bands of frequencies called sidebands (SB) on either side of the carrier frequency (C), so the filter has to pass a band of frequencies, not just a single frequency. The band of frequencies received by the receiver is called its passband (PB), and the width of the passband in kilohertz is called the bandwidth (BW). The bandwidth of the filter must be wide enough to allow the sidebands through without distortion, but narrow enough to block any interfering transmissions on adjacent frequencies (such as S2 in the diagram). The ability of the receiver to reject unwanted radio stations near in frequency to the desired station is an important parameter called selectivity determined by the filter. In modern receivers quartz crystal, ceramic resonator, or surface acoustic wave (SAW) filters are often used which have sharper selectivity compared to networks of capacitor-inductor tuned circuits.

  - **Tuning:** To select a particular station the radio is "tuned" to the frequency of the desired transmitter. The radio has a dial or digital display showing the frequency it is tuned to. *Tuning* is adjusting the frequency of the receiver's passband to the frequency of the desired radio transmitter. Turning the tuning knob changes the resonant frequency of the tuned circuit. When the resonant frequency is equal to the radio transmitter's frequency the tuned circuit oscillates in sympathy, passing the signal on to the rest of the receiver.

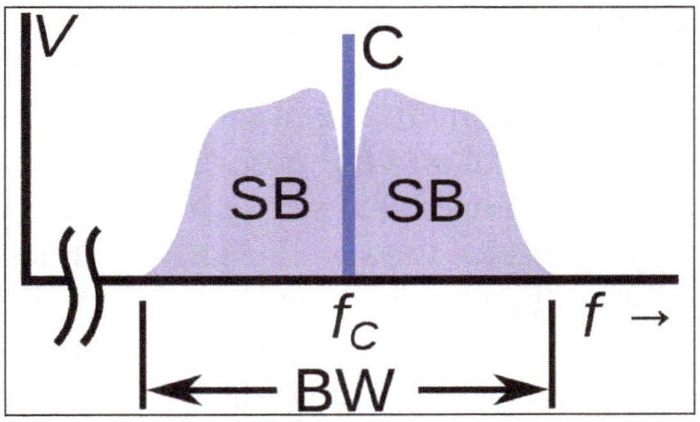

The frequency spectrum of a typical radio signal from an AM or FM radio transmitter. It consists of a strong component (C) at the carrier wave frequency f_C, with the modulation contained in narrow frequency bands called sidebands (SB) just above and below the carrier.

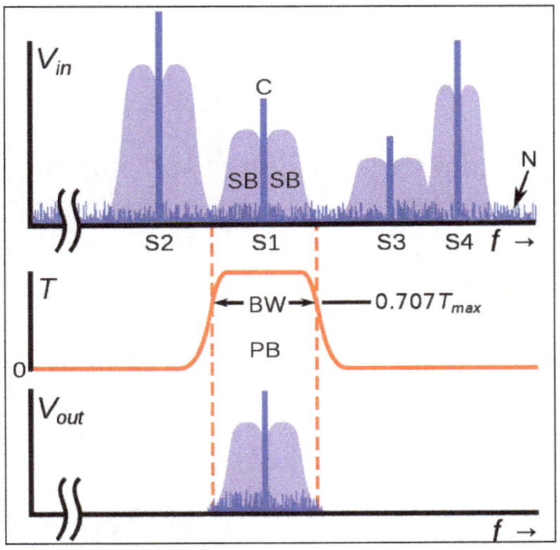

How the bandpass filter selects a single radio signal *S1* from all the radio signals received by the antenna. From top, the graphs show the voltage from the antenna applied to the filter V_{in}, the transfer function of the filter *T*, and the voltage at the output of the filter V_{out} as a function of frequency *f*. The transfer function *T* is the amount of signal that gets through the filter at each frequency:

$$V_{out}(f) = T(f)\, V_{in}(f)$$

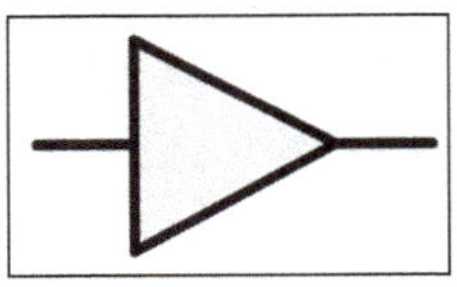

Symbol for an amplifier

- **Amplification:** The power of the radio waves picked up by a receiving antenna decreases with the square of its distance from the transmitting antenna. Even with the powerful transmitters used in radio broadcasting stations, if the receiver is more than a few miles from the transmitter the power intercepted by the receiver's antenna is very small, perhaps as low as picowatts. To increase the power of the recovered signal, an amplifier circuit uses electric power from batteries or the wall plug to increase the amplitude (voltage or current) of the signal. In most modern receivers, the electronic components which do the actual amplifying are transistors.

Receivers usually have several stages of amplification: the radio signal from the bandpass filter is amplified to make it powerful enough to drive the demodulator, then the audio signal from the demodulator is amplified to make it powerful enough to operate the speaker. The degree of amplification of a radio receiver is measured by a parameter called its *sensitivity*, which is the minimum signal strength of a station at the antenna, measured in microvolts, necessary to receive the signal clearly, with a certain signal-to-noise ratio. Since it is easy to amplify a signal to any desired degree, the limit to the sensitivity of many modern receivers is not the degree of amplification but random electronic noise present in the circuit, which can drown out a weak radio signal.

Symbol for a demodulator.

- Demodulation: After the radio signal is filtered and amplified, the receiver must extract the information-bearing modulation signal from the modulated radio frequency carrier wave. This is done by a circuit called a demodulator (detector). Each type of modulation requires a different type of demodulator
    - An AM receiver that receives an (amplitude modulated) radio signal uses an AM demodulator.
    - An FM receiver that receives a frequency modulated signal uses an FM demodulator.
    - An FSK receiver which receives frequency shift keying (used to transmit digital data in wireless devices) uses an FSK demodulator.

  Many other types of modulation are also used for specialized purposes.

The modulation signal output by the demodulator is usually amplified to increase its strength, then the information is converted back to a human-usable form by some type of transducer. An audio signal, representing sound, as in a broadcast radio, is converted to sound waves by an earphone or loudspeaker. A video signal, representing moving images, as in a television receiver, is converted to light by a display. Digital data, as in a wireless modem, is applied as input to a computer or microprocessor, which interacts with human users.

AM Demodulation

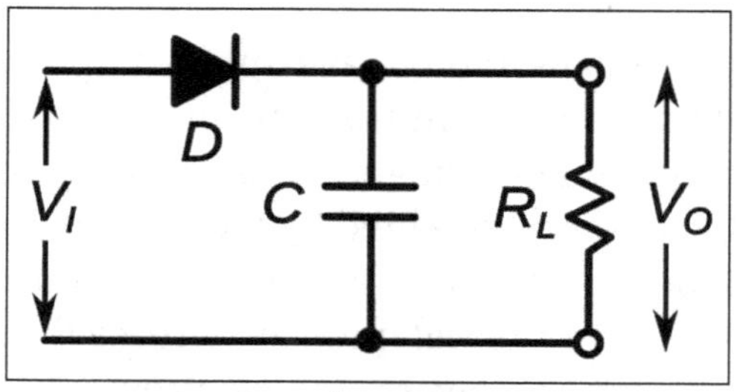

Envelope detector circuit.

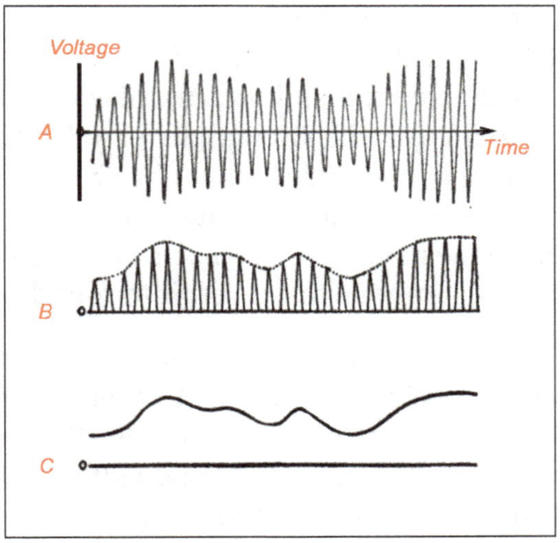

Envelope detector works.

The easiest type of demodulation to understand is AM demodulation, used in AM radios to recover the audio modulation signal, which represents sound and is converted to sound waves by the radio's speaker. It is accomplished by a circuit called an envelope detector, consisting of a diode *(D)* with a bypass capacitor *(C)* across its output.

The amplitude modulated radio signal from the tuned circuit is shown at *(A)*. The rapid oscillations are the radio frequency carrier wave. The audio signal (the sound) is contained in the slow variations (modulation) of the amplitude (size) of the waves. If it was applied directly to the speaker, this signal cannot be converted to sound, because the audio excursions are the same on both sides of the axis, averaging out to zero, which would result in no net motion of the speaker's diaphragm. *(B)* When this signal is applied as input V_I to the detector, the diode *(D)* conducts current in one direction but not in the opposite direction, thus allowing through pulses of current on only one side of the signal. In other words, it rectifies the AC current to a pulsing DC current. The resulting voltage V_O applied to the load R_L no longer averages zero; its peak value is proportional to the audio signal. *(C)* The bypass capacitor *(C)* is charged up by the current pulses from the diode, and its voltage follows the peaks of the pulses, the envelope of the audio wave. It performs a smoothing (low pass filtering) function, removing the radio frequency carrier pulses, leaving the low frequency audio signal to pass through the load R_L. The audio signal is amplified and applied to earphones or a speaker.

Tuned Radio Frequency (TRF) Receiver

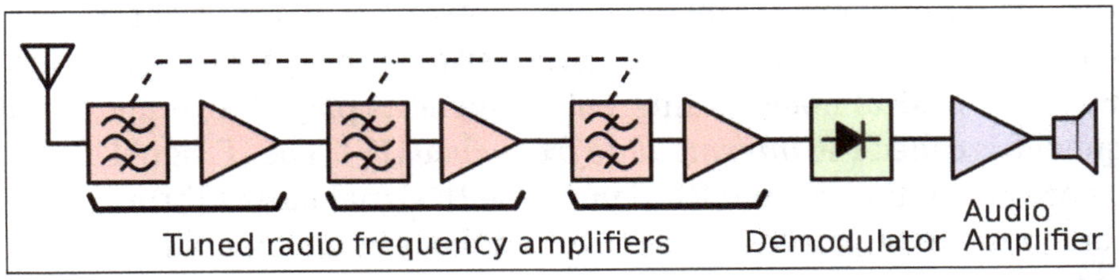

Block diagram of a tuned radio frequency receiver. To achieve enough selectivity to reject stations on adjacent frequencies, multiple cascaded bandpass filter stages had to be used. The dotted line indicates that the bandpass filters must be tuned together.

In the simplest type of radio receiver, called a tuned radio frequency (TRF) receiver, the three functions above are performed consecutively: (1) the mix of radio signals from the antenna is filtered to extract the signal of the desired transmitter; (2) this oscillating voltage is sent through a radio frequency (RF) amplifier to increase its strength to a level sufficient to drive the demodulator; (3) the demodulator recovers the modulation signal (which in broadcast receivers is an audio signal, a voltage oscillating at an audio frequency rate representing the sound waves) from the modulated radio carrier wave; (4) the modulation signal is amplified further in an audio amplifier, then is applied to a loudspeaker or earphone to convert it to sound waves.

Although the TRF receiver is used in a few applications, it has practical disadvantages which make it inferior to the superheterodyne receiver below, which is used in most applications. The drawbacks stem from the fact that in the TRF the filtering, amplification, and demodulation are done at the high frequency of the incoming radio signal. The bandwidth of a filter increases with its center frequency, so as the TRF receiver is tuned to different frequencies its bandwidth varies. Most important, the increasing congestion of the radio spectrum requires that radio channels be spaced very close together in frequency. It is extremely difficult to build filters operating at radio frequencies that have a narrow enough bandwidth to separate closely spaced radio stations. TRF receivers typically must have many cascaded tuning stages to achieve adequate selectivity. The Advantages section below describes how the superheterodyne receiver overcomes these problems.

The Superheterodyne Design

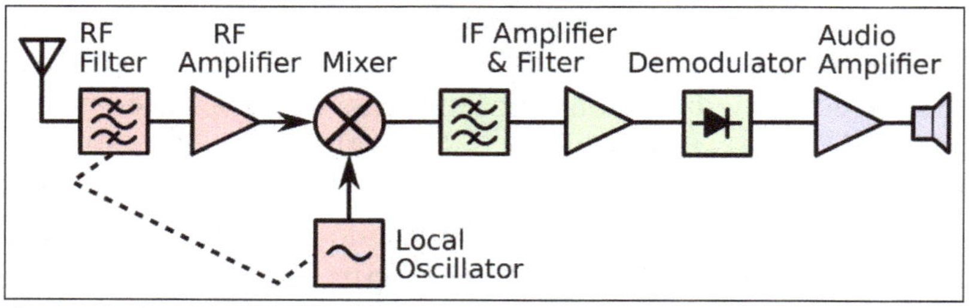

Block diagram of a superheterodyne receiver. The dotted line indicates that the RF filter and local oscillator must be tuned in tandem.

The superheterodyne receiver, invented in 1918 by Edwin Armstrong is the design used in almost all modern receivers except a few specialized applications.

In the superheterodyne, the radio frequency signal from the antenna is shifted down to a lower "intermediate frequency" (IF), before it is processed. The incoming radio frequency signal from the antenna is mixed with an unmodulated signal generated by a *local oscillator* (LO) in the receiver. The mixing is done in a nonlinear circuit called the "*mixer*". The result at the output of the mixer is a heterodyne or beat frequency at the difference between these two frequencies. The process is similar to the way two musical notes at different frequencies played together produce a beat note. This lower frequency is called the *intermediate frequency* (IF). The IF signal also has all the information that was present in the original RF signal. The IF signal passes through filter and amplifier stages, then is demodulated in a detector, recovering the original modulation.

The receiver is easy to tune; to receive a different frequency it is only necessary to change the local oscillator frequency. The stages of the receiver after the mixer operates at the fixed intermediate frequency (IF) so the IF bandpass filter does not have to be adjusted to different frequencies. The fixed frequency allows modern receivers to use sophisticated quartz crystal, ceramic resonator, or surface acoustic wave (SAW) IF filters that have very high Q factors, to improve selectivity.

The RF filter on the front end of the receiver is needed to prevent interference from any radio signals at the image frequency. Without an input filter the receiver can receive incoming RF signals at two different frequencies. The receiver can be designed to receive on either of these two frequencies; if the receiver is designed to receive on one, any other radio station or radio noise on the other frequency may pass through and interfere with the desired signal. A single tunable RF filter stage rejects the image frequency; since these are relatively far from the desired frequency, a simple filter provides adequate rejection. Rejection of interfering signals much closer in frequency to the desired signal is handled by the multiple sharply-tuned stages of the intermediate frequency amplifiers, which do no need to change their tuning. This filter does not need great selectivity, but as the receiver is tuned to different frequencies it must "track" in tandem with the local oscillator. The RF filter also serves to limit the bandwidth applied to the RF amplifier, preventing it from being overloaded by strong out-of-band signals.

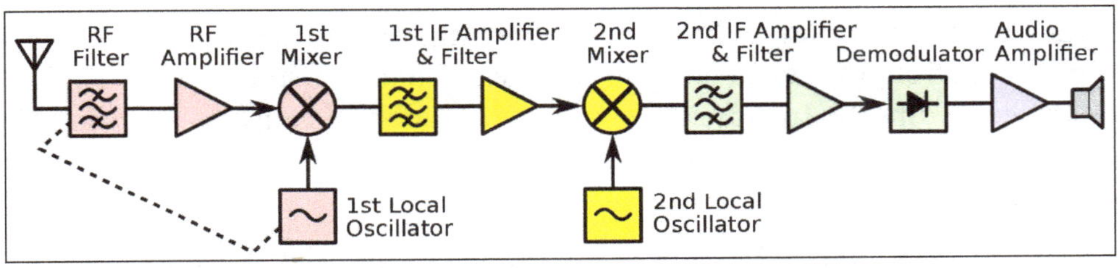

Block diagram of a dual-conversion superheterodyne receiver.

To achieve both good image rejection and selectivity, many modern superhet receivers use two intermediate frequencies; this is called a *dual-conversion* or *double-conversion* superheterodyne. The incoming RF signal is first mixed with one local oscillator signal in the first mixer to convert it to a high IF frequency, to allow efficient filtering out of the image frequency, then this first IF is mixed with a second local oscillator signal in a second mixer to convert it to a low IF frequency for good bandpass filtering. Some receivers even use triple-conversion.

At the cost of the extra stages, the superheterodyne receiver provides the advantage of greater selectivity than can be achieved with a TRF design. Where very high frequencies are in use, only the initial stage of the receiver needs to operate at the highest frequencies; the remaining stages can provide much of the receiver gain at lower frequencies which may be easier to manage. Tuning is simplified compared to a multi-stage TRF design, and only two stages need to track over the tuning range. The total amplification of the receiver is divided between three amplifiers at different frequencies; the RF, IF, and audio amplifier. This reduces problems with feedback and parasitic oscillations that are encountered in receivers where most of the amplifier stages operate at the same frequency, as in the TRF receiver.

The most important advantage is that better selectivity can be achieved by doing the filtering at the lower intermediate frequency. One of the most important parameters of a receiver is its bandwidth, the band of frequencies it accepts. In order to reject nearby interfering stations or noise, a narrow bandwidth is required. In all known filtering techniques, the bandwidth of the filter increases in proportion with the frequency, so by performing the filtering at the lower , rather than the frequency of the original radio signal , a narrower bandwidth can be achieved. Modern FM and television broadcasting, cellphones and other communications services, with their narrow channel widths, would be impossible without the superheterodyne.

Automatic Gain Control

The signal strength (amplitude) of the radio signal from a receiver's antenna varies drastically, by orders of magnitude, depending on how far away the radio transmitter is, how powerful it is, and propagation conditions along the path of the radio waves. The strength of the signal received from a given transmitter varies with time due to changing propagation conditions of the path through which the radio wave passes, such as multipath interference; this is called *fading*. In an AM receiver the amplitude of the audio signal from the detector, and the sound volume, is proportional to the amplitude of the radio signal, so fading causes variations in the volume. In addition as the receiver is tuned between strong and weak stations, the volume of the sound from the speaker would vary drastically. Without an automatic system to handle it, in an AM receiver constant adjustment of the volume control would be required.

With other types of modulation like FM or FSK the amplitude of the modulation does not vary with the radio signal strength, but in all types the demodulator requires a certain range of signal amplitude to operate properly. Insufficient signal amplitude will cause an increase of noise in the demodulator, while excessive signal amplitude will cause amplifier stages to overload (saturate), causing distortion (clipping) of the signal.

Therefore, almost all modern receivers include a feedback control system which monitors the *average* level of the radio signal at the detector, and adjusts the gain of the amplifiers to give the optimum signal level for demodulation. This is called automatic gain control (AGC). AGC can be compared to the dark adaptation mechanism in the human eye; on entering a dark room the gain of the eye is increased by the iris opening. In its simplest form an AGC system consists of a rectifier which converts the RF signal to a varying DC level, a lowpass filter to smooth the variations and produce an average level. This is applied as a control signal to an earlier amplifier stage, to control its gain. In a superheterodyne receiver AGC is usually applied to the IF amplifier, and there may be a second AGC loop to control the gain of the RF amplifier to prevent it from overloading, too.

In certain receiver designs such as modern digital receivers, a related problem is DC offset of the signal. This is corrected by a similar feedback system.

RADIO WAVE

Radio waves are a type of electromagnetic radiation with wavelengths in the electromagnetic spectrum longer than infrared light. Radio waves have frequencies as high as 300 gigahertz (GHz) to as low as 30 hertz (Hz). At 300 GHz, the corresponding wavelength is 1 mm, and at 30 Hz is 10,000 km. Like all other electromagnetic waves, radio waves travel at the speed of light in vacuum. They are generated by electric charges undergoing acceleration, such as time varying electric currents. Naturally occurring radio waves are emitted by lightning and astronomical objects.

Radio waves are generated artificially by transmitters and received by radio receivers, using antennas. Radio waves are very widely used in modern technology for fixed and mobile radio communication, broadcasting, radar and other navigation systems, communications satellites, wireless computer networks and many other applications. Different frequencies of radio waves have different propagation characteristics in the Earth's atmosphere; long waves can diffract around obstacles like mountains and follow the contour of the earth (ground waves), shorter waves can reflect off the ionosphere and return to earth beyond the horizon (skywaves), while much shorter wavelengths bend or diffract very little and travel on a line of sight, so their propagation distances are limited to the visual horizon.

To prevent interference between different users, the artificial generation and use of radio waves is strictly regulated by law, coordinated by an international body called the International Telecommunications Union (ITU), which defines radio waves as "electromagnetic waves of frequencies arbitrarily lower than 3 000 GHz, propagated in space without artificial guide". The radio spectrum is divided into a number of radio bands on the basis of frequency, allocated to different uses.

Diagram of the electric fields (E) and magnetic fields (H) of radio waves emitted by a monopole radio transmitting antenna (small dark vertical line in the center). The E and H fields are perpendicular, as implied by the phase diagram in the lower right.

Radio waves were first predicted by mathematical work done in 1867 by British mathematical physicist James Clerk Maxwell. Maxwell noticed wavelike properties of light and similarities in electrical and magnetic observations. His mathematical theory, now called Maxwell's equations, described light waves and radio waves as waves of electromagnetism that travel in space, radiated by a charged particle as it undergoes acceleration. In 1887, Heinrich Hertz demonstrated the reality of Maxwell's electromagnetic waves by experimentally generating radio waves in his laboratory, showing that they exhibited the same wave properties as light: standing waves, refraction, diffraction, and polarization. Radio waves, originally called "Hertzian waves", were first used for communication in the mid 1890s by Guglielmo Marconi, who developed the first practical radio transmitters and receivers. The modern term "radio wave" replaced the original name "Hertzian wave" around 1912.

Speed, Wavelength and Frequency

Animated diagram of a half-wave dipole antenna receiving a radio wave. The antenna consists of two metal rods connected to a receiver R. The electric field (E, green arrows) of the incoming wave pushes the electrons in the rods back and forth, charging the ends alternately positive (+) and negative (−). Since the length of the antenna is one half the wavelength of the wave, the oscillating field induces standing waves of voltage (V, represented by red band) and current in the rods. The oscillating currents (black arrows) flow down the transmission line and through the receiver (represented by the resistance R).

Radio waves in a vacuum travel at the speed of light. When passing through a material medium, they are slowed according to that object's permeability and permittivity. Air is thin enough that in the Earth's atmosphere radio waves travel very close to the speed of light.

The wavelength is the distance from one peak of the wave's electric field (wave's peak/crest) to the next, and is inversely proportional to the frequency of the wave. The distance a radio wave travels in one second, in a vacuum, is 299,792,458 meters (983,571,056 ft) which is the wavelength of a 1 hertz radio signal. A 1 megahertz radio signal has a wavelength of 299.8 meters (984 ft).

Propagation

The study of radio propagation, how radio waves move in free space and over the surface of the Earth, is vitally important in the design of practical radio systems. Radio waves passing through different environments experience reflection, refraction, polarization, diffraction, and absorption. Different frequencies experience different combinations of these phenomena in the Earth's atmosphere, making certain radio bands more useful for specific purposes than others. Practical radio systems mainly use three different techniques of radio propagation to communicate:

- Line of sight: This refers to radio waves that travel in a straight line from the transmitting antenna to the receiving antenna. It does not necessarily require a cleared sight path; at lower frequencies radio waves can pass through buildings, foliage and other obstructions. This is the only method of propagation possible at frequencies above 30 MHz. On the surface of the Earth, line of sight propagation is limited by the visual horizon to about 64 km (40 mi). This is the method used by cell phones, FM, television broadcasting and radar. By using dish antennas to transmit beams of microwaves, point-to-point microwave relay links transmit telephone and television signals over long distances up to the visual horizon. Ground stations can communicate with satellites and spacecraft billions of miles from Earth.

    - Indirect propagation: Radio waves can reach points beyond the line-of-sight by diffraction and reflection. Diffraction allows a radio wave to bend around obstructions such as a building edge, a vehicle, or a turn in a hall. Radio waves also partially reflect from surfaces such as walls, floors, ceilings, vehicles and the ground. These propagation methods occur in short range radio communication systems such as cell phones, cordless phones, walkie-talkies, and wireless networks. A drawback of this mode is multipath propagation, in which radio waves travel from the transmitting to the receiving antenna via multiple paths. The waves interfere, often causing fading and other reception problems.

- Ground waves: At lower frequencies below 2 MHz, in the medium wave and longwave bands, due to diffraction vertically polarized radio waves can bend over hills and mountains, and propagate beyond the horizon, traveling as surface waves which follow the contour of the Earth. This allows mediumwave and longwave broadcasting stations to have coverage areas beyond the horizon, out to hundreds of miles. As the frequency drops, the losses decrease and the achievable range increases. Military very low frequency (VLF) and extremely low frequency (ELF) communication systems can communicate over most of the Earth, and with submarines hundreds of feet underwater.

- Skywaves: At medium wave and shortwave wavelengths, radio waves reflect off conductive layers of charged particles (ions) in a part of the atmosphere called the ionosphere. So radio waves directed at an angle into the sky can return to Earth beyond the horizon; this is called "skip" or "skywave" propagation. By using multiple skips communication at intercontinental distances can be achieved. Skywave propagation is variable and dependent on atmospheric conditions; it is most reliable at night and in the winter. Widely used during the first half of the 20th century, due to its unreliability skywave communication has mostly been abandoned. Remaining uses are by military over-the-horizon (OTH) radar systems, by some automated systems, by radio amateurs, and by shortwave broadcasting stations to broadcast to other countries.

Radio Communication

In radio communication systems, information is carried across space using radio waves. At the sending end, the information to be sent, in the form of a time-varying electrical signal, is applied to a radio transmitter. The information signal can be an audio signal representing sound from a microphone, a video signal representing moving images from a video camera, or a digital signal representing data from a computer. In the transmitter, an electronic oscillator generates an alternating current oscillating at a radio frequency, called the *carrier wave* because it serves to "carry" the information through the air. The information signal is used to modulate the carrier, altering some aspect of it, "piggybacking" the information on the carrier. The modulated carrier is amplified and applied to an antenna. The oscillating current pushes the electrons in the antenna back and forth, creating oscillating electric and magnetic fields, which radiate the energy away from the antenna as radio waves. The radio waves carry the information to the receiver location.

At the receiver, the oscillating electric and magnetic fields of the incoming radio wave push the electrons in the receiving antenna back and forth, creating a tiny oscillating voltage which is a weaker replica of the current in the transmitting antenna. This voltage is applied to the radio receiver, which extracts the information signal. The receiver first uses a bandpass filter to separate the desired radio station's radio signal from all the other radio signals picked up by the antenna, then amplifies the signal so it is stronger, then finally extracts the information-bearing modulation signal in a demodulator. The recovered signal is sent to a loudspeaker or earphone to produce sound, or a television display screen to produce a visible image, or other devices. A digital data signal is applied to a computer or microprocessor, which interacts with a human user.

The radio waves from many transmitters pass through the air simultaneously without interfering with each other. They can be separated in the receiver because each transmitter's radio waves oscillate at a different rate, in other words each transmitter has a different frequency, measured in kilohertz (kHz), megahertz (MHz) or gigahertz (GHz). The bandpass filter in the receiver consists of a tuned circuit which acts like a resonator, similarly to a tuning fork. It has a natural resonant frequency at which it oscillates. The resonant frequency is set equal to the frequency of the desired radio station. The oscillating radio signal from the desired station causes the tuned circuit to oscillate in sympathy, and it passes the signal on to the rest of the receiver. Radio signals at other frequencies are blocked by the tuned circuit and not passed on.

Biological and Environmental Effects

Radio waves are *nonionizing radiation*, which means they do not have enough energy to separate electrons from atoms or molecules, ionizing them, or break chemical bonds, causing chemical reactions or DNA damage. The main effect of absorption of radio waves by materials is to heat them, similarly to the infrared waves radiated by sources of heat such as a space heater or wood fire. The oscillating electric field of the wave causes polar molecules to vibrate back and forth, increasing the temperature; this is how a microwave oven cooks food. However, unlike infrared waves, which are mainly absorbed at the surface of objects and cause surface heating, radio waves are able to penetrate the surface and deposit their energy inside materials and biological tissues. The depth to which radio waves penetrate decreases with their frequency, and also depends on the material's resistivity and permittivity; it is given by a parameter called the *skin depth* of the material, which is the depth within which 63% of the energy is deposited. For example, the 2.45 GHz radio waves (microwaves) in a microwave oven penetrate most foods approximately 2.5 to 3.8 cm (1 to 1.5 inches). Radio waves have been applied to the body for 100 years in the medical therapy of diathermy for deep heating of body tissue, to promote increased blood flow and healing. More recently they have been used to create higher temperatures in hyperthermia treatment and to kill cancer cells. Looking into a source of radio waves at close range, such as the waveguide of a working radio transmitter, can cause damage to the lens of the eye by heating. A strong enough beam of radio waves can penetrate the eye and heat the lens enough to cause cataracts.

Since the heating effect is in principle no different from other sources of heat, most research into possible health hazards of exposure to radio waves has focused on "nonthermal" effects; whether radio waves have any effect on tissues besides that caused by heating. Electromagnetic radiation has been classified by the International Agency for Research on Cancer (IARC) as "Possibly carcinogenic to humans". The conceivable evidence of cancer risk via Personal exposure to RF-EMF with mobile telephone use was identified.

Radio waves can be shielded against by a conductive metal sheet or screen, an enclosure of sheet or screen is called a Faraday cage. A metal screen shields against radio waves as well as a solid sheet as long as the holes in the screen are smaller than about 1/20 of wavelength of the waves.

Measurement

Since radio frequency radiation has both an electric and a magnetic component, it is often convenient to express intensity of radiation field in terms of units specific to each component. The unit *volts per meter* (V/m) is used for the electric component, and the unit *amperes per meter* (A/m) is used for the magnetic component. One can speak of an electromagnetic field, and these units are used to provide information about the levels of electric and magnetic field strength at a measurement location.

Another commonly used unit for characterizing an RF electromagnetic field is *power density*. Power density is most accurately used when the point of measurement is far enough away from the RF emitter to be located in what is referred to as the far field zone of the radiation pattern. In closer proximity to the transmitter, i.e., in the "near field" zone, the physical relationships between the electric and magnetic components of the field can be complex, and it is best to use the field strength units discussed above. Power density is measured in terms of power per unit area, for example, milliwatts per square centimeter (mW/cm^2). When speaking of frequencies in the microwave range and higher, power density is usually used to express intensity since exposures that might occur would likely be in the far field zone.

TRANSMISSION LINE

In radio-frequency engineering, a transmission line is a specialized cable or other structure designed to conduct alternating current of radio frequency, that is, currents with a frequency high enough that their wave nature must be taken into account. Transmission lines are used for purposes such as connecting radio transmitters and receivers with their antennas (they are then called feed lines or feeders), distributing cable television signals, trunklines routing calls between telephone switching centres, computer network connections and high speed computer data buses.

Ordinary electrical cables suffice to carry low frequency alternating current (AC), such as mains power, which reverses direction 100 to 120 times per second, and audio signals. However, they cannot be used to carry currents in the radio frequency range, above about 30 kHz, because the energy tends to radiate off the cable as radio waves, causing power losses. Radio frequency currents also tend to reflect from discontinuities in the cable such as connectors and joints, and travel back down the cable toward the source. These reflections act as bottlenecks, preventing the signal power from reaching the destination. Transmission lines use specialized construction, and impedance matching, to carry electromagnetic signals with minimal reflections and power losses. The distinguishing feature of most transmission lines is that they have uniform cross sectional dimensions along their length, giving them a uniform *impedance*, called the characteristic impedance, to prevent reflections. Types of transmission line include parallel line (ladder line, twisted pair), coaxial cable, and planar transmission lines such as stripline and microstrip. The higher the frequency of electromagnetic waves moving through a given cable or medium, the shorter the wavelength of the waves. Transmission lines become necessary when the transmitted frequency's wavelength is sufficiently short that the length of the cable becomes a significant part of a wavelength.

At microwave frequencies and above, power losses in transmission lines become excessive, and waveguides are used instead, which function as "pipes" to confine and guide the electromagnetic waves. Some sources define waveguides as a type of transmission line. At even higher frequencies, in the terahertz, infrared and visible ranges, waveguides in turn become lossy, and optical methods, (such as lenses and mirrors), are used to guide electromagnetic waves.

The theory of sound wave propagation is very similar mathematically to that of electromagnetic waves, so techniques from transmission line theory are also used to build structures to conduct acoustic waves; and these are called acoustic transmission lines.

Applicability

In many electric circuits, the length of the wires connecting the components can for the most part be ignored. That is, the voltage on the wire at a given time can be assumed to be the same at all points. However, when the voltage changes in a time interval comparable to the time it takes for the signal to travel down the wire, the length becomes important and the wire must be treated as a transmission line. Stated another way, the length of the wire is important when the signal includes frequency components with corresponding wavelengths comparable to or less than the length of the wire.

A common rule of thumb is that the cable or wire should be treated as a transmission line if the length is greater than 1/10 of the wavelength. At this length the phase delay and the interference of any reflections on the line become important and can lead to unpredictable behaviour in systems which have not been carefully designed using transmission line theory.

The Four Terminal Model

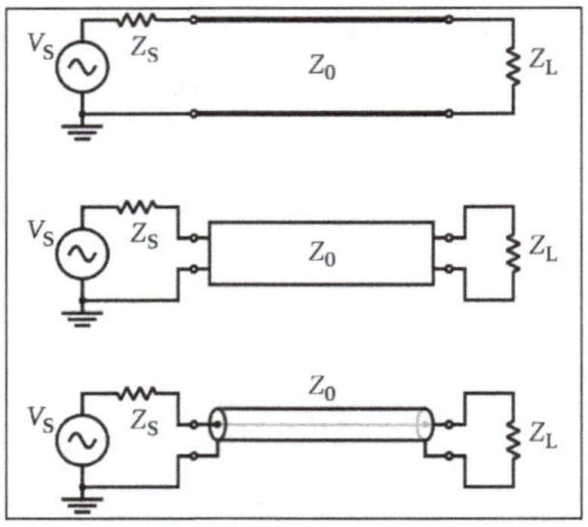

Variations on the schematic electronic symbol for a transmission line.

For the purposes of analysis, an electrical transmission line can be modelled as a two-port network (also called a quadripole), as follows:

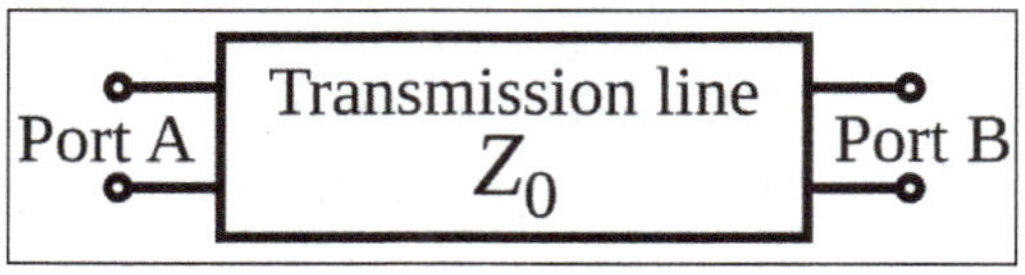

In the simplest case, the network is assumed to be linear (i.e. the complex voltage across either port is proportional to the complex current flowing into it when there are no reflections), and the two ports are assumed to be interchangeable. If the transmission line is uniform along its length, then its behaviour is largely described by a single parameter called the *characteristic impedance*, symbol Z_o. This is the ratio of the complex voltage of a given wave to the complex current of the same wave at any point on the line. Typical values of Z_o are 50 or 75 ohms for a coaxial cable, about 100 ohms for a twisted pair of wires, and about 300 ohms for a common type of untwisted pair used in radio transmission.

When sending power down a transmission line, it is usually desirable that as much power as possible will be absorbed by the load and as little as possible will be reflected back to the source. This can be ensured by making the load impedance equal to Z_o, in which case the transmission line is said to be *matched.*

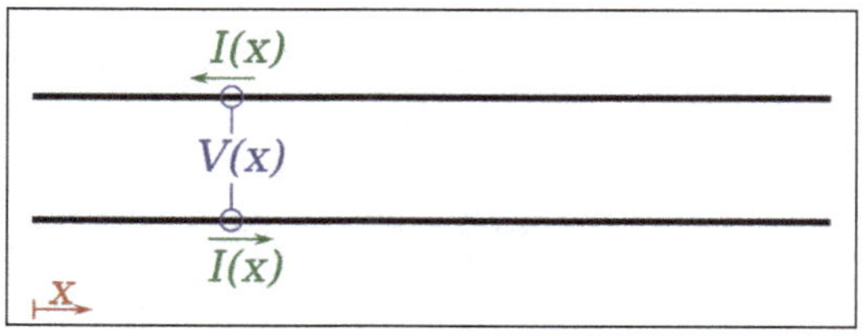

A transmission line is drawn as two black wires. At a distance *x* into the line, there is current *I(x)* travelling through each wire, and there is a voltage difference *V(x)* between the wires. If the current and voltage come from a single wave (with no reflection), then $V(x) / I(x) = Z_o$, where Z_o is the *characteristic impedance* of the line.

Some of the power that is fed into a transmission line is lost because of its resistance. This effect is called *ohmic* or *resistive* loss. At high frequencies, another effect called *dielectric loss* becomes significant, adding to the losses caused by resistance. Dielectric loss is caused when the insulating material inside the transmission line absorbs energy from the alternating electric field and converts it to heat. The transmission line is modelled with a resistance (R) and inductance (L) in series with a capacitance (C) and conductance (G) in parallel. The resistance and conductance contribute to the loss in a transmission line.

The total loss of power in a transmission line is often specified in decibels per metre (dB/m), and usually depends on the frequency of the signal. The manufacturer often supplies a chart showing the loss in dB/m at a range of frequencies. A loss of 3 dB corresponds approximately to a halving of the power.

High-frequency transmission lines can be defined as those designed to carry electromagnetic waves whose wavelengths are shorter than or comparable to the length of the line. Under these conditions, the approximations useful for calculations at lower frequencies are no longer accurate. This often occurs with radio, microwave and optical signals, metal mesh optical filters, and with the signals found in high-speed digital circuits.

Telegrapher's Equations

The telegrapher's equations (or just telegraph equations) are a pair of linear differential equations which describe the voltage (V) and current (I) on an electrical transmission line with distance and time. They were developed by Oliver Heaviside who created the *transmission line model*, and are based on Maxwell's Equations.

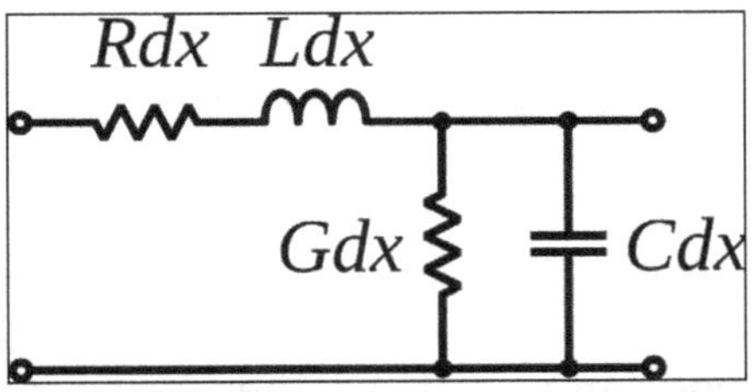

Schematic representation of the elementary component of a transmission line.

The transmission line model is an example of the distributed element model. It represents the transmission line as an infinite series of two-port elementary components, each representing an infinitesimally short segment of the transmission line:

- The distributed resistance R of the conductors is represented by a series resistor (expressed in ohms per unit length).
- The distributed inductance L (due to the magnetic field around the wires, self-inductance, etc.) is represented by a series inductor (in henries per unit length).
- The capacitance C between the two conductors is represented by a shunt capacitor (in farads per unit length).
- The conductance G of the dielectric material separating the two conductors is represented by a shunt resistor between the signal wire and the return wire (in siemens per unit length).

The model consists of an *infinite series* of the elements shown in the figure, and the values of the components are specified *per unit length* so the picture of the component can be misleading. R, L, C, and G may also be functions of frequency. An alternative notation is to use R, L, C and G to emphasize that the values are derivatives with respect to length. These quantities can also be known as the primary line constants to distinguish from the secondary line constants derived from them, these being the propagation constant, attenuation constant and phase constant.

The line voltage $V(x)$ and the current $I(x)$ can be expressed in the frequency domain as:

$$\frac{\partial V(x)}{\partial x} = -(R + j\omega L)I(x)$$

$$\frac{\partial I(x)}{\partial x} = -(G + j\omega C)V(x).$$

Special Case of a Lossless Line

When the elements L and C are negligibly small the transmission line is considered as a lossless structure. In this hypothetical case, the model depends only on the R and G elements which greatly simplifies the analysis. For a lossless transmission line, the second order steady-state Telegrapher's equations are:

$$\frac{\partial^2 V(x)}{\partial x^2} + \omega^2 LCV(x) = 0$$

$$\frac{\partial^2 I(x)}{\partial x^2} + \omega^2 LCI(x) = 0.$$

These are wave equations which have plane waves with equal propagation speed in the forward and reverse directions as solutions. The physical significance of this is that electromagnetic waves propagate down transmission lines and in general, there is a reflected component that interferes with the original signal. These equations are fundamental to transmission line theory.

General Case of a Line with Losses

In the general case the loss terms, R and G, are both included, and the full form of the Telegrapher's equations become:

$$\frac{\partial^2 V(x)}{\partial x^2} = \gamma^2 V(x)$$

$$\frac{\partial^2 I(x)}{\partial x^2} = \gamma^2 I(x)$$

where γ is the (complex) propagation constant. These equations are fundamental to transmission line theory. They are also wave equations, and have solutions similar to the special case, but which are a mixture of sines and cosines with exponential decay factors. Solving for the propagation constant γ in terms of the primary parameters R, L, G, and C gives:

$$\gamma = \sqrt{(R + j\omega L)(G + j\omega C)}$$

and the characteristic impedance can be expressed as:

$$Z_0 = \sqrt{\frac{R + j\omega L}{G + j\omega C}}.$$

The solutions for $V(x)$ and $I(x)$ are:

$$V(x) = V_{(+)}e^{-\gamma x} + V_{(-)}e^{+\gamma x}$$

$$I(x) = \frac{1}{Z_0}\left(V_{(+)}e^{-\gamma x} - V_{(-)}e^{+\gamma x}\right).$$

The constants $V_{(\pm)}$ must be determined from boundary conditions. For a voltage pulse $V_{in}(t)$, starting at $x = 0$ and moving in the positive x direction, then the transmitted pulse $V_{out}(x,t)$ at position x can be obtained by computing the Fourier Transform, $\tilde{V}(\omega)$, of $V_{in}(t)$, attenuating each frequency component by $e^{-\mathrm{Re}(\gamma)x}$, advancing its phase by $-\mathrm{Im}(\gamma)x$, and taking the inverse Fourier Transform. The real and imaginary parts of γ can be computed as,

$$\mathrm{Re}(\gamma) = \alpha = (a^2 + b^2)^{1/4}\cos(\psi)$$

$$\mathrm{Im}(\gamma) = \beta = (a^2 + b^2)^{1/4}\sin(\psi)$$

with,

$$a \equiv RG - \omega^2 LC = \omega^2 LC\left[\left(\frac{R}{\omega L}\right)\left(\frac{G}{\omega C}\right) - 1\right]$$

$$b \equiv \omega CR + \omega LG = \omega^2 LC\left(\frac{R}{\omega L} + \frac{G}{\omega C}\right)$$

the right-hand expressions holding when neither L, nor C, nor ω is zero, and with,

$$\psi \equiv \tfrac{1}{2}\operatorname{atan} 2(b,a)$$

where atan2 is the everywhere-defined form of two-parameter arctangent function, with arbitrary value zero when both arguments are zero.

Special and Low Loss Case

For small losses and high frequencies, the general equations can be simplified: If $\frac{R}{\omega L} \ll 1$ and $\frac{G}{\omega C} \ll 1$ then,

$$\mathrm{Re}(\gamma) = \alpha \approx \tfrac{1}{2}\sqrt{LC}\left(\frac{R}{L}+\frac{G}{C}\right)$$

$$\mathrm{Im}(\gamma) = \beta \approx \omega\sqrt{LC}\ .$$

Noting that an advance in phase by $-\omega\delta$ is equivalent to a time delay by δ, $V_{out}(t)$ can be simply computed as:

$$V_{\mathrm{out}}(x,t) \approx V_{\mathrm{in}}(t-\sqrt{LC}\,x)e^{-\frac{1}{2}\sqrt{LC}\left(\frac{R}{L}+\frac{G}{C}\right)x}.$$

Heaviside Condition

The Heaviside condition is a special case where the wave travels down the line without any dispersion distortion. The condition for this to take place is:

$$\frac{G}{C}=\frac{R}{L}$$

Input Impedance of Transmission Line

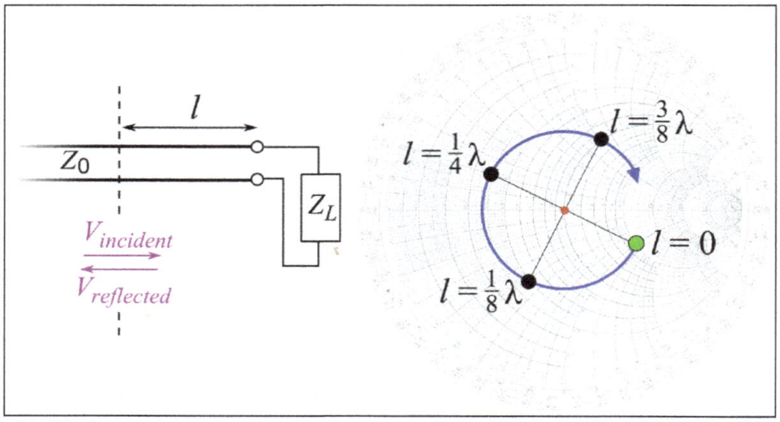

Looking towards a load through a length ℓ of lossless transmission line, the impedance changes as ℓ increases, following the blue circle on this impedance Smith chart. (This impedance is characterized by its reflection coefficient, which is the reflected voltage divided by the incident voltage.) The blue circle, centred within the chart, is sometimes called an *SWR circle* (short for *constant standing wave ratio*).

The characteristic impedance Z_0 of a transmission line is the ratio of the amplitude of a *single* voltage wave to its current wave. Since most transmission lines also have a reflected wave, the characteristic impedance is generally not the impedance that is measured on the line.

The impedance measured at a given distance ℓ from the load impedance Z_L may be expressed as:

$$Z_{\mathrm{in}}(\ell) = \frac{V(\ell)}{I(\ell)} = Z_0\frac{1+\Gamma_L e^{-2\gamma\ell}}{1-\Gamma_L e^{-2\gamma\ell}},$$

where γ is the propagation constant and $\Gamma_L = \frac{Z_L - Z_0}{Z_L + Z_0}$ is the voltage reflection coefficient measured at the load end of the transmission line. Alternatively, the above formula can be rearranged to express the input impedance in terms of the load impedance rather than the load voltage reflection coefficient:

$$Z_{in}(\ell) = Z_0 \frac{Z_L + Z_0 \tanh(\gamma \ell)}{Z_0 + Z_L \tanh(\gamma \ell)}.$$

Input Impedance of Lossless Transmission Line

For a lossless transmission line, the propagation constant is purely imaginary, $\gamma = j\beta$, so the above formulas can be rewritten as,

$$Z_{in}(\ell) = Z_0 \frac{Z_L + jZ_0 \tan(\beta \ell)}{Z_0 + jZ_L \tan(\beta \ell)}$$

where $\beta = \frac{2\pi}{\lambda}$ is the wavenumber.

In calculating β ,the wavelength is generally different *inside* the transmission line to what it would be in free-space. Consequently, the velocity constant of the material the transmission line is made of needs to be taken into account when doing such a calculation.

Special Cases of Lossless Transmission Lines

Half Wave Length

For the special case where $\beta \ell = n\pi$ where n is an integer (meaning that the length of the line is a multiple of half a wavelength), the expression reduces to the load impedance so that,

$$Z_{in} = Z_L$$

for all n. This includes the case when $n = 0$, meaning that the length of the transmission line is negligibly small compared to the wavelength. The physical significance of this is that the transmission line can be ignored (i.e. treated as a wire) in either case.

Quarter Wave Length

For the case where the length of the line is one quarter wavelength long, or an odd multiple of a quarter wavelength long, the input impedance becomes:

$$Z_{in} = \frac{Z_0^2}{Z_L}.$$

Matched Load

Another special case is when the load impedance is equal to the characteristic impedance of the line (i.e. the line is *matched*), in which case the impedance reduces to the characteristic impedance of the line so that,

$$Z_{\mathrm{in}} = Z_{\mathrm{L}} = Z_0$$

for all ℓ and all λ.

Short

For the case of a shorted load (i.e. $Z_{\mathrm{L}} = 0$), the input impedance is purely imaginary and a periodic function of position and wavelength (frequency):

$$Z_{\mathrm{in}}(\ell) = jZ_0 \tan(\beta\ell).$$

Open

For the case of an open load (i.e. $Z_{\mathrm{L}} = \infty$), the input impedance is once again imaginary and periodic:

$$Z_{\mathrm{in}}(\ell) = -jZ_0 \cot(\beta\ell).$$

Stepped Transmission Line

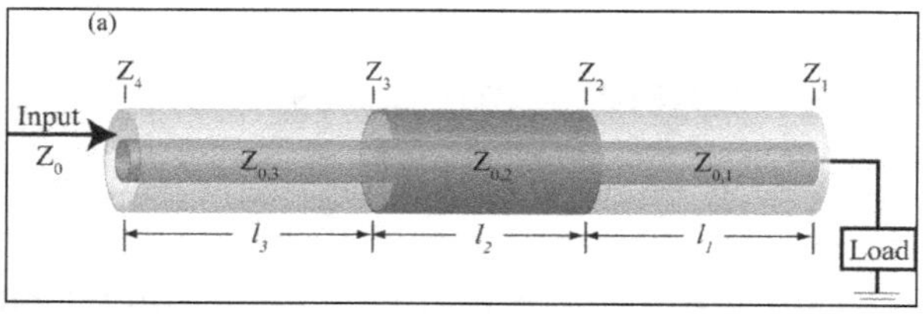

A simple example of stepped transmission line consisting of three segments.

A stepped transmission line is used for broad range impedance matching. It can be considered as multiple transmission line segments connected in series, with the characteristic impedance of each individual element to be $Z_{0,\mathrm{i}}$. The input impedance can be obtained from the successive application of the chain relation:

$$Z_{\mathrm{i}+1} = Z_{0,\mathrm{i}} \frac{Z_{\mathrm{i}} + jZ_{0,\mathrm{i}} \tan(\beta_{\mathrm{i}}\ell_{\mathrm{i}})}{Z_{0,\mathrm{i}} + jZ_{\mathrm{i}} \tan(\beta_{\mathrm{i}}\ell_{\mathrm{i}})}$$

where β_{i} is the wave number of the i -th transmission line segment and ℓ_{i} is the length of this segment, and Z_{i} is the front-end impedance that loads the i -th segment.

Because the characteristic impedance of each transmission line segment $Z_{0,\mathrm{i}}$ is often different from that of the input cable Z_0, the impedance transformation circle is off-centred along the x axis of the Smith Chart whose impedance representation is usually normalized against Z_0.

The stepped transmission line is an example of a distributed element circuit. A large variety of other circuits can also be constructed with transmission lines including filters, power dividers and directional couplers.

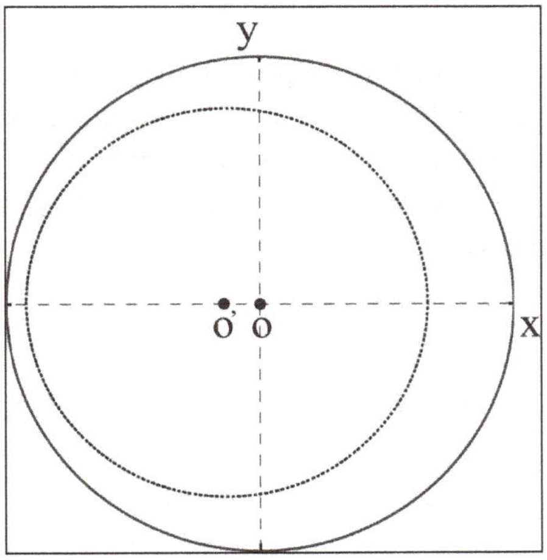

The impedance transformation circle along a transmission line whose characteristic impedance $Z_{0,i}$ is smaller than that of the input cable Z_0. And as a result, the impedance curve is off-centred towards the $-x$ axis. Conversely, if $Z_{0,i} > Z_0$, the impedance curve should be off-centred towards the $+x$ axis.

Practical Types

Microstrip

A type of transmission line called a cage line, used for high power, low frequency applications. It functions similarly to a large coaxial cable. This example is the antenna feed line for a longwave radio transmitter in Poland, which operates at a frequency of 225 kHz and a power of 1200 kW.

A microstrip circuit uses a thin flat conductor which is parallel to a ground plane. Microstrip can be made by having a strip of copper on one side of a printed circuit board (PCB) or ceramic substrate while the other side is a continuous ground plane. The width of the strip, the thickness of the insulating layer (PCB or ceramic) and the dielectric constant of the insulating layer determine the characteristic impedance. Microstrip is an open structure whereas coaxial cable is a closed structure.

Stripline

A stripline circuit uses a flat strip of metal which is sandwiched between two parallel ground planes. The insulating material of the substrate forms a dielectric. The width of the strip, the thickness of the substrate and the relative permittivity of the substrate determine the characteristic impedance of the strip which is a transmission line.

Coplanar Waveguide

A coplanar waveguide consists of a center strip and two adjacent outer conductors, all three of them flat structures that are deposited onto the same insulating substrate and thus are located in the same plane ("coplanar"). The width of the center conductor, the distance between inner and outer conductors, and the relative permittivity of the substrate determine the characteristic impedance of the coplanar transmission line.

Balanced Lines

A balanced line is a transmission line consisting of two conductors of the same type, and equal impedance to ground and other circuits. There are many formats of balanced lines, amongst the most common are twisted pair, star quad and twin-lead.

Twisted Pair

Twisted pairs are commonly used for terrestrial telephone communications. In such cables, many pairs are grouped together in a single cable, from two to several thousand. The format is also used for data network distribution inside buildings, but the cable is more expensive because the transmission line parameters are tightly controlled.

Star Quad

Star quad is a four-conductor cable in which all four conductors are twisted together around the cable axis. It is sometimes used for two circuits, such as 4-wire telephony and other telecommunications applications. In this configuration each pair uses two non-adjacent conductors. Other times it is used for a single, balanced line, such as audio applications and 2-wire telephony. In this configuration two non-adjacent conductors are terminated together at both ends of the cable, and the other two conductors are also terminated together.

When used for two circuits, crosstalk is reduced relative to cables with two separate twisted pairs.

When used for a single, balanced line, magnetic interference picked up by the cable arrives as a virtually perfect common mode signal, which is easily removed by coupling transformers.

The combined benefits of twisting, balanced signalling, and quadrupole pattern give outstanding noise immunity, especially advantageous for low signal level applications such as microphone cables, even when installed very close to a power cable. The disadvantage is that star quad, in combining two conductors, typically has double the capacitance of similar two-conductor twisted and shielded audio cable. High capacitance causes increasing distortion and greater loss of high frequencies as distance increases.

Twin-lead

Twin-lead consists of a pair of conductors held apart by a continuous insulator. By holding the conductors a known distance apart, the geometry is fixed and the line characteristics are reliably consistent. It is lower loss than coaxial cable because the characteristic impedance of twin-lead is generally higher than coaxial cable, leading to lower resistive losses due to the reduced current. However, it is more susceptible to interference.

Lecher Lines

Lecher lines are a form of parallel conductor that can be used at UHF for creating resonant circuits. They are a convenient practical format that fills the gap between lumped components (used at HF/VHF) and resonant cavities (used at UHF/SHF).

Single-wire Line

Unbalanced lines were formerly much used for telegraph transmission, but this form of communication has now fallen into disuse. Cables are similar to twisted pair in that many cores are bundled into the same cable but only one conductor is provided per circuit and there is no twisting. All the circuits on the same route use a common path for the return current (earth return). There is a power transmission version of single-wire earth return in use in many locations.

General Applications

Signal Transfer

Electrical transmission lines are very widely used to transmit high frequency signals over long or short distances with minimum power loss. One familiar example is the down lead from a TV or radio aerial to the receiver.

Pulse Generation

Transmission lines are also used as pulse generators. By charging the transmission line and then discharging it into a resistive load, a rectangular pulse equal in length to twice the electrical length of the line can be obtained, although with half the voltage. A Blumlein transmission line is a related pulse forming device that overcomes this limitation. These are sometimes used as the pulsed power sources for radar transmitters and other devices.

Stub Filters

If a short-circuited or open-circuited transmission line is wired in parallel with a line used to transfer signals from point A to point B, then it will function as a filter. The method for making stubs is similar to the method for using Lecher lines for crude frequency measurement, but it is 'working backwards'. One method recommended in the RSGB's radiocommunication handbook is to take an open-circuited length of transmission line wired in parallel with the feeder delivering signals from an aerial. By cutting the free end of the transmission line, a minimum in the strength of the signal observed at a receiver can be found. At this stage the stub filter will reject this frequency and the odd harmonics, but if the free end of the stub is shorted then the stub will become a filter rejecting the even harmonics.

Coaxial Cable

A type of copper cable specially built with a metal shield and other components engineered to block signal interference. It is primarily used by cable TV companies to connect their satellite antenna facilities to customer homes and businesses. It is also sometimes used by telephone companies to connect central offices to telephone poles near customers. Some homes and offices use coaxial cable, too, but its widespread use as an Ethernet connectivity medium in enterprises and data centers has been supplanted by the deployment of twisted pair cabling.

Coaxial cable received its name because it includes one physical channel that carries the signal surrounded after a layer of insulation by another concentric physical channel, both running along the same axis. The outer channel serves as a ground. Many of these cables or pairs of coaxial tubes can be placed in a single outer sheathing and, with repeaters, can carry information for a great distance.

It was invented in 1880 by English engineer and mathematician Oliver Heaviside, who patented the invention and design that same year. AT&T established its first cross-continental coaxial transmission system in 1940. Depending on the carrier technology used and other factors, twisted pair copper wire and optical fiber are alternatives to coaxial cable.

Coaxial cables have concentric layers of electrical conductors and insulating material. This construction ensures signals are enclosed within the cable and prevents electrical noise from interfering with the signal.

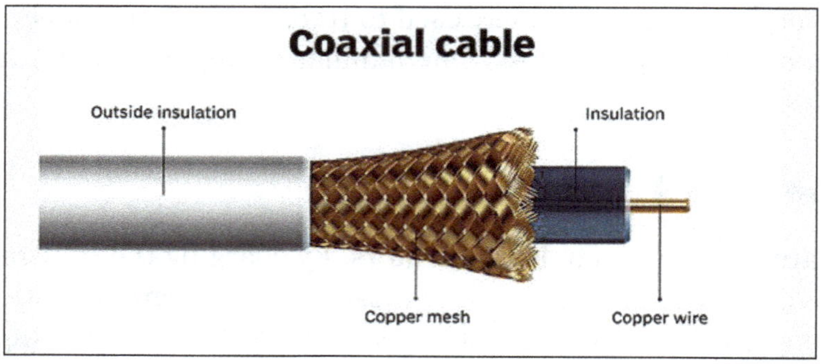

The center conductor layer is a thin conducting wire, either solid or braided copper. A dielectric layer, made up of an insulating material with very well-defined electrical characteristics, surrounds the wire. A shield layer then surrounds the dielectric layer with metal foil or braided copper mesh. The whole assembly is wrapped in an insulating jacket. The outer metal shield layer of the coaxial cable is typically grounded in the connectors at both ends to shield the signals and as a place for stray interference signals to dissipate.

A key to coaxial cable design is a tight control of cable dimensions and materials. Together, they ensure the characteristic impedance of the cable takes on a fixed value. High-frequency signals are partially reflected at impedance mismatches, causing errors.

Characteristic impedance is sensitive to signal frequency. Above 1 GHz, the cable maker must use a dielectric that does not attenuate the signal too much or change the characteristic impedance in a way that creates signal reflections.

Electrical characteristics of coax are application-dependent and crucial for good performance. Two standard characteristic impedances are 50 ohms, used in moderate power environments, and 75 ohms, common for connections to antennas and residential installations.

Types of Coaxial Cables

There are numerous types of coaxial cables, some types include:

- Hard-line coaxial cable: Which relies on round copper tubing and a combination of metals as a shield, such as aluminum or copper. These cables are commonly used to connect a transmitter to an antenna.
- Triaxial cable: Which has a third layer of shielding that is grounded to protect signals transmitted down the cable.
- Rigid-line coaxial cables: Which are made up of twin copper tubes that function as unbendable pipes. These lines are designed for indoor use between high-power radio frequency (RF) transmitters.
- Radiating cable: Which mimics many components of the hard-line cable, but with tuned slots in the shielding matched to the RF wavelength at which the cable will operate. It is commonly used in elevators, military equipment and underground tunnels.

Types of Connectors

There are many different types of coaxial cable connectors separated by two styles—male and female connectors. Connector types include:

- BNC: Standing for Bayonet Neil-Concelman, this connector is used with television, video signal and radio below a frequency of 4GHz.
- TNC: Standing for Threaded Neil-Concelman, this connector is a threaded version of the BNC connector and is used in cellphones. TNC connectors operate up to 12 GHz.
- SMA: Standing for SubMiniature version A, this connector is used with cellphones, Wi-Fi antenna systems, microwave systems and radios. SMA connectors operate up to 18GHz.
- SMB: Standing for Subminiature version B, this connector may be used with telecommunications hardware.
- QMA: QMA connectors are a quick-locking variant of SMA connectors used with industrial and communications hardware.
- RCA: Standing for Radio Corporation of America, these are connectors used in audio and video. These are the grouped yellow, white and red cables used with older televisions. RCA connectors are also called A/V jacks.
- F connectors: Also called F-types, these are used in digital and cable televisions. These commonly use RG6 or RG 59 cables.

Uses of Coaxial Cables

In the home and small offices, short coaxial cables are used for cable television, home video equipment, amateur radio equipment and measuring devices. Historically, coaxial cables were also used as an early form of Ethernet, supporting speeds of up to 10 Mbps, but coax has supplanted by the use of twisted pair cabling. However, they remain widely in use for cable broadband internet. Coaxial cables are also used in automobiles, aircraft, military and medical equipment, as well as to connect satellite dishes, radio and television antennae to their respective receivers.

Common coax cable types and uses		
Cable Type	Ω	Use
RG-6	75	Video, TV
RG-11	75	Long runs
RG-59	75	Video, TV

Standards

Most coaxial specifications have an impedance of 50, 52, 75 or 93 ohms. Because of widespread use in the cable television industry, RG-6 cables with double or quad shields and impedance of 75 ohms have become a de facto standard for many industries. Nearly 50 distinct standards exist for coaxial cable, often designed for specific use cases in amateur radio or low-loss cable television. Other examples include RG-59/U used for carrying broadband signal from closed-circuit TV systems or RG-214/U used for high-frequency signal transmission.

Connectors for coax range from simple single connectors used on cable TV systems to complicated combinations of multiple thin coax links, mixed with power and other signal connections, housed in semi-custom bodies. These are commonly found in military electronics and avionics.

Mechanical stiffness can vary tremendously, depending on the internal construction and intended use of the coaxial cabling. For example, high-power cables are often made with thick insulation and are very stiff.

Some cables are deliberately made with thick center wires, resulting in skin-effect resistance. It results from high-frequency signals traveling on the surface of the conductor, not throughout. If the center conductor is larger, it results in a stiff cable with low loss per meter.

Interference Issues

Coaxial cables can experience a variety of different forms of interference. Signal leakage occurs when the electromagnetic field passes through the shielding on the outside of the cable. In other cases, an outside signal can leak through the insulation. Straight-line feeds to commercial radio broadcast towers have the least leakage and interference because these cables have smooth, conductive shields with few gaps in them. Interference is most significant in nuclear reactors, where special shielding is needed.

Difference between RG59 and RG6

RG59 and RG6 cables are commonly used in satellite television and cable modems. Older installations used the RG59 cable before the implantation of the RG6 cable. The RG59 cable is thinner at a 20 American Wire Gauge (AWG) and has a copper center conductor. This cable is more likely to be found in older buildings and is better for CCTV and analog video systems.

The RG6 cable is a larger 18 AWG cable and also has a copper center conductor. The RG6 cable is used with high-bandwidth and high-frequency hardware, where internet and satellite signals can run at a higher frequency compared to traditional analog video.

FREE-SPACE OPTICAL COMMUNICATION

Free-space optical communication (FSO) is an optical communication technology that uses light propagating in free space to wirelessly transmit data for telecommunications or computer networking. "Free space" means air, outer space, vacuum, or something similar. This contrasts with using solids such as optical fiber cable.

The technology is useful where the physical connections are impractical due to high costs or other considerations.

Usage and Technologies

Free-space point-to-point optical links can be implemented using infrared laser light, although low-data-rate communication over short distances is possible using LEDs. Infrared Data Association (IrDA) technology is a very simple form of free-space optical communications. On the communications side the FSO technology is considered as a part of the optical wireless communications applications. Free-space optics can be used for communications between spacecraft.

Commercial Products

- In 2008, MRV Communications introduced a free-space optics (FSO)-based system with a data rate of 10 Gbit/s initially claiming a distance of 2 km at high availability. This equipment is no longer available; before end-of-life, the product's useful distance was changed down to 350 m.
- In 2013, the company MOSTCOM started to serially produce a new wireless communication system that also had a data rate of 10 Gbit/s as well as an improved range of up to 2.5 km, but to get to 99.99% uptime the designers used an RF hybrid solution, meaning the data rate drops to extremely low levels during atmospheric disturbances (typically down to 10 Mbit/s). In April 2014, the company with Scientific and Technological Centre "Fiord" demonstrated the transmission speed 30 Gbit/s under "laboratory conditions".
- LightPointe offers many similar hybrid solutions to MOSTCOM's offering.

Useful Distances

The reliability of FSO units has always been a problem for commercial telecommunications. Consistently, studies find too many dropped packets and signal errors over small ranges (400 to 500 meters). This is from both independent studies, such as in the Czech republic, as well as formal internal nationwide studies, such as one conducted by MRV FSO staff. Military based studies consistently produce longer estimates for reliability, projecting the maximum range for terrestrial links is of the order of 2 to 3 km (1.2 to 1.9 mi). All studies agree the stability and quality of the link is highly dependent on atmospheric factors such as rain, fog, dust and heat. Relays may be employed to extend the range for FSO communications.

Extending the useful Distance

The main reason terrestrial communications have been limited to non-commercial telecommunications functions is fog. Fog consistently keeps FSO laser links over 500 meters from achieving a year-round bit error rate of 1 per 100,000. Several entities are continually attempting to overcome these key disadvantages to FSO communications and field a system with a better quality of service. DARPA has sponsored over US$130 million in research towards this effort, with the ORCA and ORCLE programs.

Other non-government groups are fielding tests to evaluate different technologies that some claim have the ability to address key FSO adoption challenges. As of October 2014, none have fielded a working system that addresses the most common atmospheric events.

FSO research from 1998–2006 in the private sector totaled $407.1 million, divided primarily among four start-up companies. All four failed to deliver products that would meet telecommunications quality and distance standards:

- Terabeam received approximately $575 million in funding from investors such as Softbank, Mobius Venture Capital and Oakhill Venture Partners. AT&T and Lucent backed this attempt. The work ultimately failed, and the company was purchased in 2004 for $52 million (excluding warrants and options) by Falls Church, Va.-based YDI, effective June 22, 2004, and used the name Terabeam for the new entity. On September 4, 2007, Terabeam (then headquartered in San Jose, California) announced it would change its name to Proxim Wireless Corporation, and change its NASDAQ stock symbol from TRBM to PRXM.

- AirFiber received $96.1 million in funding, and never solved the weather issue. They sold out to MRV communications in 2003, and MRV sold their FSO units until 2012 when the end-of-life was abruptly announced for the Terescope series.

- LightPointe Communications received $76 million in start-up funds, and eventually reorganized to sell hybrid FSO-RF units to overcome the weather-based challenges.
- The Maxima Corporation published its operating theory in Science (magazine), and received $9 million in funding before permanently shutting down. No known spin-off or purchase followed this effort.
- Wireless Excellence developed and launched CableFree UNITY solutions that combine FSO with millimeter wave and radio technologies to extend distance, capacity and availability, with a goal of making FSO a more useful and practical technology.

One private company published a paper on November 20, 2014, claiming they had achieved commercial reliability (99.999% availability) in extreme fog. There is no indication this product is currently commercially available.

Extraterrestrial

The massive advantages of laser communication in space have multiple space agencies racing to develop a stable space communication platform, with many significant demonstrations and achievements.

Operational Systems

The first gigabit laser-based communication was achieved by the European Space Agency and called the European Data Relay System (EDRS) on November 28, 2014. The system is operational and is being used on a daily basis.

Demonstrations

NASA's OPALS announced a breakthrough in space-to-ground communication December 9, 2014, uploading 175 megabytes in 3.5 seconds. Their system is also able to re-acquire tracking after the signal was lost due to cloud cover.

In the early morning hours of Oct. 18, NASA's Lunar Laser Communication Demonstration (LLCD) made history, transmitting data from lunar orbit to Earth at a rate of 622 Megabits-per-second (Mbps). LLCD was flown aboard the Lunar Atmosphere and Dust Environment Explorer satellite (LADEE), whose primary science mission was to investigate the tenuous and exotic atmosphere that exists around the moon.

In January 2013, NASA used lasers to beam an image of the Mona Lisa to the Lunar Reconnaissance Orbiter roughly 390,000 km (240,000 mi) away. To compensate for atmospheric interference, an error correction code algorithm similar to that used in CDs was implemented.

A two-way distance record for communication was set by the Mercury laser altimeter instrument aboard the MESSENGER spacecraft, and was able to communicate across a distance of 24 million km (15 million miles), as the craft neared Earth on a fly-by in May, 2005. The previous record had been set with a one-way detection of laser light from Earth, by the Galileo probe, of 6 million km in 1992. Quote from Laser Communication in Space Demonstrations (EDRS).

Commercial Use

Various planned satellite constellations such as SpaceX's Starlink intended to provide global broadband coverage employ laser communication for inter-satellite links between the several hundred to thousand satellites effectively creating a space-based optical mesh network.

LEDs

RONJA is a free implementation of FSO using high-intensity LEDs.

In 2001, Twibright Labs released Ronja Metropolis, an open source DIY 10 Mbit/s full duplex LED FSO over 1.4 km In 2004, a Visible Light Communication Consortium was formed in Japan. This was based on work from researchers that used a white LED-based space lighting system for indoor local area network (LAN) communications. These systems present advantages over traditional UHF RF-based systems from improved isolation between systems, the size and cost of receivers/ transmitters, RF licensing laws and by combining space lighting and communication into the same system. In January 2009, a task force for visible light communication was formed by the Institute of Electrical and Electronics Engineers working group for wireless personal area network standards known as IEEE 802.15.7. A trial was announced in 2010, in St. Cloud, Minnesota.

Amateur radio operators have achieved significantly farther distances using incoherent sources of light from high-intensity LEDs. One reported 173 miles (278 km) in 2007. However, physical limitations of the equipment used limited bandwidths to about 4 kHz. The high sensitivities required of the detector to cover such distances made the internal capacitance of the photodiode used a dominant factor in the high-impedance amplifier which followed it, thus naturally forming a low-pass filter with a cut-off frequency in the 4 kHz range. Use of lasers can reach very high data rates which are comparable to fiber communications.

Projected data rates and future data rate claims vary. A low-cost white LED (GaN-phosphor) which could be used for space lighting can typically be modulated up to 20 MHz. Data rates of over 100 Mbit/s can be easily achieved using efficient modulation schemes and Siemens claimed to have achieved over 500 Mbit/s in 2010. Research published in 2009, used a similar system for traffic control of automated vehicles with LED traffic lights.

In September 2013, pureLiFi, the Edinburgh start-up working on Li-Fi, also demonstrated high speed point-to-point connectivity using any off-the-shelf LED light bulb. In previous work, high bandwidth specialist LEDs have been used to achieve the high data rates. The new system, the

Li-1st, maximizes the available optical bandwidth for any LED device, thereby reducing the cost and improving the performance of deploying indoor FSO systems.

Engineering Details

Typically, best use scenarios for this technology are:

- LAN-to-LAN connections on campuses at Fast Ethernet or Gigabit Ethernet speeds.
- LAN-to-LAN connections in a city, a metropolitan area network.
- To cross a public road or other barriers which the sender and receiver do not own.
- Speedy service delivery of high-bandwidth access to optical fiber networks.
- Converged Voice-Data-Connection.
- Temporary network installation (for events or other purposes).
- Reestablish high-speed connection quickly (disaster recovery).
- As an alternative or upgrade add-on to existing wireless technologies.
    - Especially powerful in combination with auto aiming systems, this way you could power moving cars or you can power your laptop while you move or use auto-aiming nodes to create a network with other nodes.
- As a safety add-on for important fiber connections (redundancy).
- For communications between spacecraft, including elements of a satellite constellation.
- For inter- and intra-chip communication.

The light beam can be very narrow, which makes FSO hard to intercept, improving security. In any case, it is comparatively easy to encrypt any data traveling across the FSO connection for additional security. FSO provides vastly improved electromagnetic interference (EMI) behavior compared to using microwaves.

Technical Advantages

- Ease of deployment.
- Can be used to power devices.
- License-free long-range operation (in contrast with radio communication).
- High bit rates.
- Low bit error rates.
- Immunity to electromagnetic interference.
- Full duplex operation.

- Protocol transparency.
- Increased security when working with narrow beam(s).
- No Fresnel zone necessary.
- Reference open source implementation.

Range Limiting Factors

For terrestrial applications, the principal limiting factors are:

- Fog (10 to ~100 dB/km attenuation).
- Beam dispersion.
- Atmospheric absorption.
- Rain.
- Snow.
- Terrestrial scintillation.
- Interference from background light sources (including the sun).
- Shadowing.
- Pointing stability in wind.
- Pollution/smog.

These factors cause an attenuated receiver signal and lead to higher bit error ratio (BER). To overcome these issues, vendors found some solutions, like multi-beam or multi-path architectures, which use more than one sender and more than one receiver. Some state-of-the-art devices also have larger fade margin (extra power, reserved for rain, smog, fog). To keep an eye-safe environment, good FSO systems have a limited laser power density and support laser classes 1 or 1M. Atmospheric and fog attenuation, which are exponential in nature, limit practical range of FSO devices to several kilometres. However the free space optics, based on 1550 nm wavelength, have considerably lower optical loss than free space optics, using 830 nm wavelength, in dense fog conditions. FSO using wavelength 1550 nm system are capable of transmitting several times higher power than systems with 850 nm and are at the same time safe to the human eye (1M class). Additionally, some free space optics, such as EC SYSTEM, ensure higher connection reliability in bad weather conditions by constantly monitoring link quality to regulate laser diode transmission power with built-in automatic gain control.

FIBER-OPTIC COMMUNICATION

For gigabits and beyond gigabits transmission of data, the fiber optic communication is the ideal choice. This type of communication is used to transmit voice, video, telemetry and data over long distances and local area networks or computer networks. A fiber Optic Communication System uses light wave technology to transmit the data over a fiber by changing electronic signals into light.

Some exceptional characteristic features of this type of communication system like large bandwidth, smaller diameter, light weight, long distance signal transmission, low attenuation, transmission security, and so on make this communication a major building block in any telecommunication infrastructure. The subsequent information on fiber optic communication system highlights its characteristic features, basic elements and other details.

Fiber optic communication.

Unlike copper wire based transmission where the transmission entirely depends on electrical signals passing through the cable, the fiber optics transmission involves transmission of signals in the form of light from one point to the other. Furthermore, a fiber optic communication network consists of transmitting and receiving circuitry, a light source and detector devices like the ones shown in the figure.

When the input data, in the form of electrical signals, is given to the transmitter circuitry, it converts them into light signal with the help of a light source. This source is of LED whose amplitude, frequency and phases must remain stable and free from fluctuation in order to have efficient transmission. The light beam from the source is carried by a fiber optic cable to the destination circuitry wherein the information is transmitted back to the electrical signal by a receiver circuit.

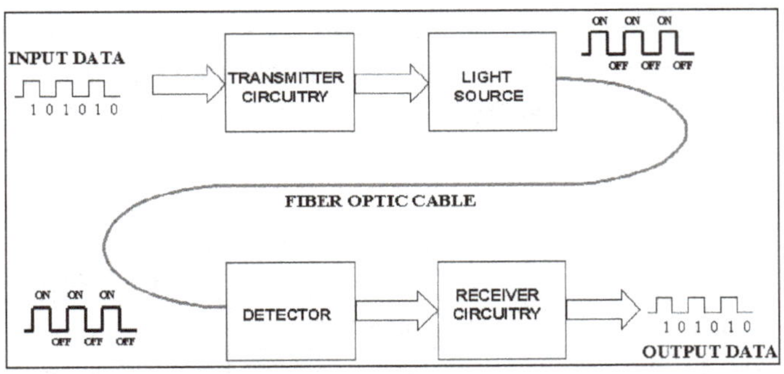

Working of Fiber optic communication.

The Receiver circuit consists of a photo detector along with an appropriate electronic circuit, which is capable of measuring magnitude, frequency and phase of the optic field. This type of communication uses the wave lengths near to the infrared band that are just above the visible range. Both LED and Laser can be used as light sources based on the application.

Elements of a Fiber Optic Communication System

There are three main basic elements of fiber optic communication system. They are

- Compact Light Source.
- Low loss Optical Fiber.
- Photo Detector.

Accessories like connectors, switches, couplers, multiplexing devices, amplifiers and splices are also essential elements in this communication system.

Compact Light Source

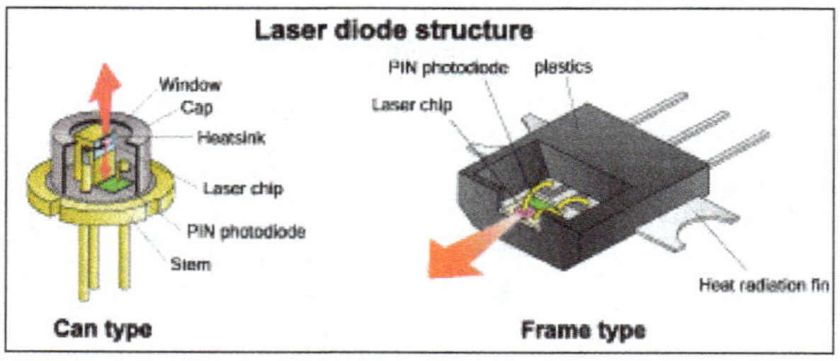

Laser Diodes.

Depending on the applications like local area networks and the long haul communication systems, the light source requirements vary. The requirements of the sources include power, speed, spectral line width, noise, ruggedness, cost, temperature, and so on. Two components are used as light sources: light emitting diodes (LED's) and laser diodes.

The light emitting diodes are used for short distances and low data rate applications due to their low bandwidth and power capabilities. Two such LEDs structures include Surface and Edge Emitting Systems. The surface emitting diodes are simple in design and are reliable, but due to its broader line width and modulation frequency limitation edge emitting diode are mostly used. Edge emitting diodes have high power and narrower line width capabilities.

For longer distances and high data rate transmission, Laser Diodes are preferred due to its high power, high speed and narrower spectral line width characteristics. But these are inherently non-linear and more sensitive to temperature variations.

LED versus Laser

Characteristic	LED	Laser
Output power	Lower	Higher
Spectral width	Wider	Narrower
Numerical aperture	Larger	Smaller
Speed	Slower	Faster
Cost	Less	More
Ease of operation	Easier	More difficult

Nowadays many improvements and advancements have made these sources more reliable. Both these sources are modulated using either direct or external modulation techniques.

Low Loss Optical Fiber

Optical fiber is a cable, which is also known as cylindrical dielectric waveguide made of low loss material. An optical fiber also considers the parameters like the environment in which it is operating, the tensile strength, durability and rigidity. The Fiber optic cable is made of high quality extruded glass (si) or plastic, and it is flexible. The diameter of the fiber optic cable is in between 0.25 to 0.5mm (slightly thicker than a human hair).

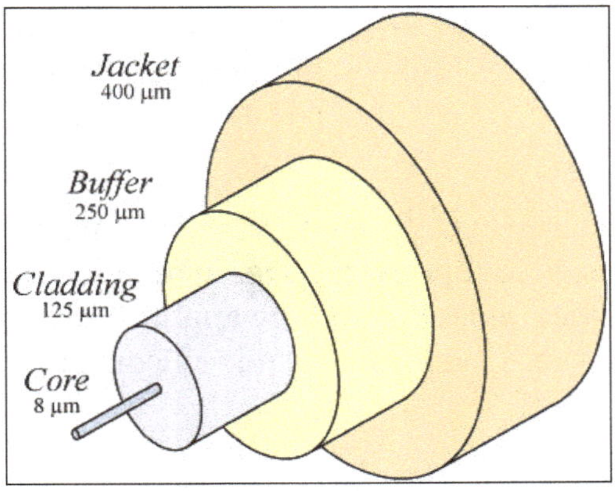

Fiber Optic Cable.

A Fiber Optic Cable consists of four parts:

- Core
- Cladding
- Buffer
- Jacket
- Core

The core of a fiber cable is a cylinder of plastic that runs all along the fiber cable's length, and offers protection by cladding. The diameter of the core depends on the application used. Due to internal reflection, the light travelling within the core reflects from the core, the cladding boundary. The core cross section needs to be a circular one for most of the applications.

Cladding

Cladding is an outer optical material that protects the core. The main function of the cladding is that it reflects the light back into the core. When light enters through the core (dense material) into the cladding(less dense material), it changes its angle, and then reflects back to the core.

Buffer

The main function of the buffer is to protect the fiber from damage and thousands of optical fibers arranged in hundreds of optical cables. These bundles are protected by the cable's outer covering that is called jacket.

Jacket

Fiber optic cable's jackets are available in different colors that can easily make us recognize the exact color of the cable we are dealing with. The color yellow clearly signifies a single mode cable, and orange color indicates multimode.

Types of Optical Fibers

Single-Mode Fibers: Single mode fibers are used to transmit one signal per fiber; these fibers are used in telephone and television sets. Single mode fibers have small cores.

Multi-Mode Fibers: Multimode fibers are used to transmit many signals per fiber; these signals are used in computer and local area networks that have larger cores.

Photo Detectors

The purpose of photo detectors is to convert the light signal back to an electrical signal. Two types of photo detectors are mainly used for optical receiver in optical communication system: PN photo diode and avalanche photo diode. Depending on the application's wavelengths, the material composition of these devices vary. These materials include silicon, germanium, InGaAs, etc.

References

- Telecommunications-media, topic: britannica.com, Retrieved 9 January, 2019
- Stephenson, Parks (November 2001). "The Marconi Wireless Installation in R.M.S. Titanic". Old Timer›s Bulletin. 42 (4). Retrieved May 22, 2016. Copied on Stephenson›s marconigraph.com personal website
- Coaxial-cable-illustrated, definition: searchnetworking.techtarget.com, Retrieved 10 February, 2019
- Serway, Raymond; Faughn, Jerry; Vuille, Chris (2008). College Physics, 8th Ed. Cengage Learning. P. 714. ISBN 0495386936
- Basic-elements-of-fiber-optic-communication-system-and-its-working: elprocus.com, Retrieved 11 March, 2019
- "Agents Classified by the IARC Monographs, Volumes 1–123". Monographs.iarc.fr. IARC. 9 Nov 2018. Retrieved 9 Jan 2019

3

Network Topology and Switching

Network Topology can be defined as the layout of the network which connects different nodes of a network to establish a connection. The hardware device that channels incoming data from multiple input ports to a specific output port that will take it towards its final destination is known as a network switch. All the varied aspects of network topology and switching have been carefully analyzed in this chapter.

NETWORK TOPOLOGY

A network topology is the arrangement of nodes - usually switches, routers, or software switch/router features - and connections in a network, often represented as a graph. The topology of the network, and the relative locations of the source and destination of traffic flows on the network, determine the optimum path for each flow and the extent to which redundant options for routing exist in the event of a failure. There are two ways of defining network geometry: the physical topology and the logical (or signal) topology.

The physical topology of a network is the layout of nodes and physical connections, including wires (Ethernet, DSL), fiber optics, and microwave.

Types of Network Topology

The different types of network topologies are:

Bus Topology

Bus topology is a network type in which every computer and network device is connected to single cable. When it has exactly two endpoints, then it is called Linear Bus topology.

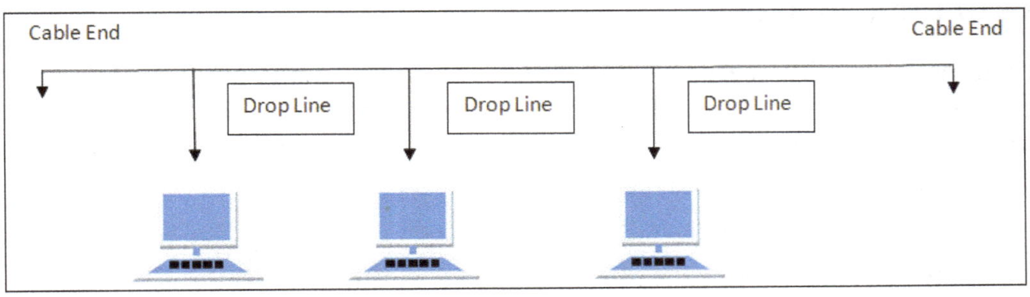

Features of Bus Topology

- It transmits data only in one direction.
- Every device is connected to a single cable.

Advantages of Bus Topology

- It is cost effective.
- Cable required is least compared to other network topology.
- Used in small networks.
- It is easy to understand.
- Easy to expand joining two cables together.

Disadvantages of Bus Topology

- Cables fails then whole network fails.
- If network traffic is heavy or nodes are more the performance of the network decreases.
- Cable has a limited length.
- It is slower than the ring topology.

RING Topology

It is called ring topology because it forms a ring as each computer is connected to another computer, with the last one connected to the first. Exactly two neighbours for each device.

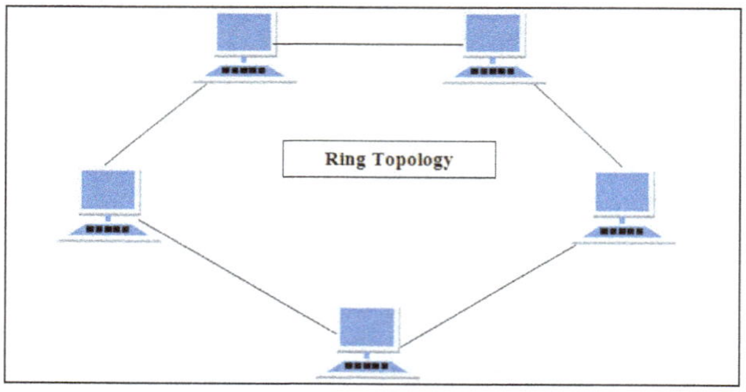

Features of Ring Topology

- A number of repeaters are used for Ring topology with large number of nodes, because if someone wants to send some data to the last node in the ring topology with 100 nodes, then the data will have to pass through 99 nodes to reach the 100th node. Hence to prevent data loss repeaters are used in the network.
- The transmission is unidirectional, but it can be made bidirectional by having 2 connections between each Network Node, it is called Dual Ring Topology.

- In Dual Ring Topology, two ring networks are formed, and data flow is in opposite direction in them. Also, if one ring fails, the second ring can act as a backup, to keep the network up.
- Data is transferred in a sequential manner that is bit by bit. Data transmitted, has to pass through each node of the network, till the destination node.

Advantages of Ring Topology

- Transmitting network is not affected by high traffic or by adding more nodes, as only the nodes having tokens can transmit data.
- Cheap to install and expand.

Disadvantages of Ring Topology

- Troubleshooting is difficult in ring topology.
- Adding or deleting the computers disturbs the network activity.
- Failure of one computer disturbs the whole network.

Star Topology

In this type of topology all the computers are connected to a single hub through a cable. This hub is the central node and all others nodes are connected to the central node.

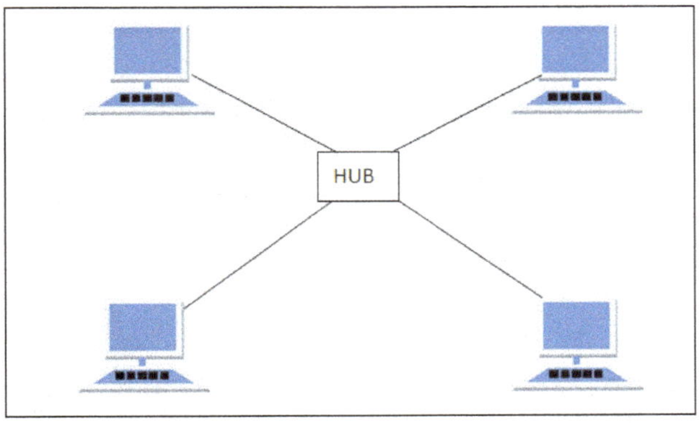

Features of Star Topology

- Every node has its own dedicated connection to the hub.
- Hub acts as a repeater for data flow.
- Can be used with twisted pair, Optical Fibre or coaxial cable.

Advantages of Star Topology

- Fast performance with few nodes and low network traffic.
- Hub can be upgraded easily.

- Easy to troubleshoot.
- Easy to setup and modify.
- Only that node is affected which has failed, rest of the nodes can work smoothly.

Disadvantages of Star Topology

- Cost of installation is high.
- Expensive to use.
- If the hub fails then the whole network is stopped because all the nodes depend on the hub.
- Performance is based on the hub that is it depends on its capacity.

Mesh Topology

It is a point-to-point connection to other nodes or devices. All the network nodes are connected to each other. Mesh has n(n-1)/2 physical channels to link n devices.

There are two techniques to transmit data over the Mesh topology, they are :

- Routing,
- Flooding.

Routing

In routing, the nodes have a routing logic, as per the network requirements. Like routing logic to direct the data to reach the destination using the shortest distance. Or, routing logic which has information about the broken links, and it avoids those node etc. We can even have routing logic, to re-configure the failed nodes.

Flooding

In flooding, the same data is transmitted to all the network nodes, hence no routing logic is required. The network is robust, and the its very unlikely to lose the data. But it leads to unwanted load over the network.

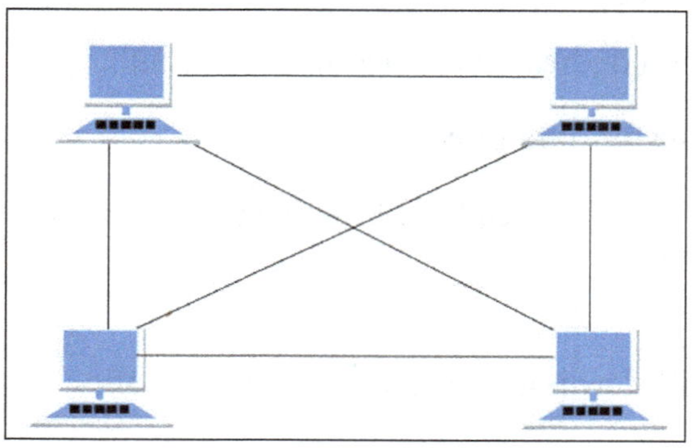

Types of Mesh Topology

Partial Mesh Topology : In this topology some of the systems are connected in the same fashion as mesh topology but some devices are only connected to two or three devices.

Full Mesh Topology : Each and every nodes or devices are connected to each other.

Features of Mesh Topology

- Fully connected.
- Robust.
- Not flexible.

Advantages of Mesh Topology

- Each connection can carry its own data load.
- It is robust.
- Fault is diagnosed easily.
- Provides security and privacy.

Disadvantages of Mesh Topology

- Installation and configuration is difficult.
- Cabling cost is more.
- Bulk wiring is required.

TREE Topology

It has a root node and all other nodes are connected to it forming a hierarchy. It is also called hierarchical topology. It should at least have three levels to the hierarchy.

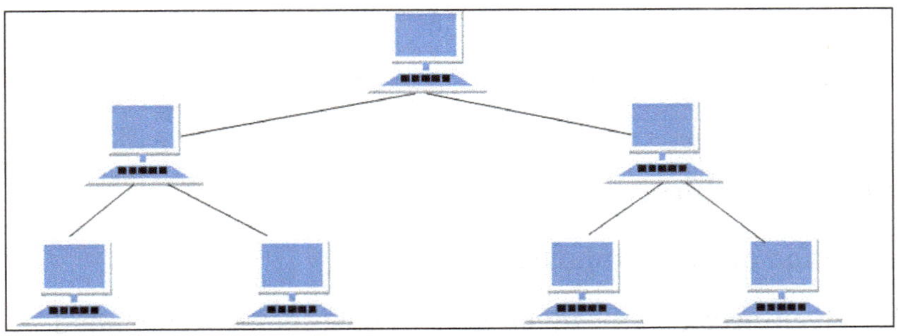

Features of Tree Topology

- Ideal if workstations are located in groups.
- Used in Wide Area Network.

Advantages of Tree Topology

- Extension of bus and star topologies.
- Expansion of nodes is possible and easy.
- Easily managed and maintained.
- Error detection is easily done.

Disadvantages of Tree Topology

- Heavily cabled.
- Costly.
- If more nodes are added maintenance is difficult.
- Central hub fails, network fails.

Hybrid Topology

It is two different types of topologies which is a mixture of two or more topologies. For example if in an office in one department ring topology is used and in another star topology is used, connecting these topologies will result in Hybrid Topology (ring topology and star topology).

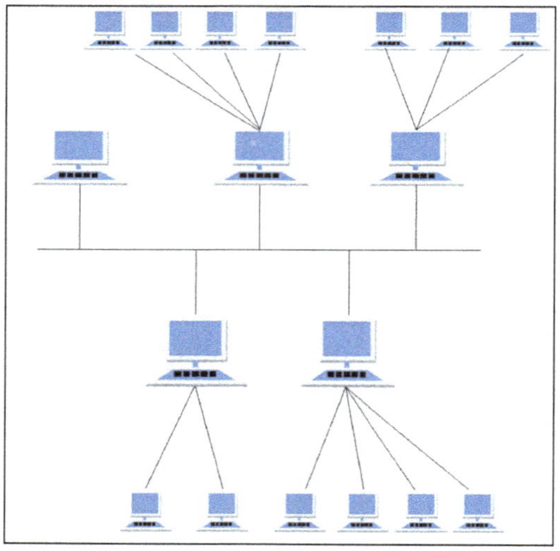

Features of Hybrid Topology

- It is a combination of two or topologies.
- Inherits the advantages and disadvantages of the topologies included.

Advantages of Hybrid Topology

- Reliable as Error detecting and trouble shooting is easy.
- Effective.

- Scalable as size can be increased easily.
- Flexible.

Disadvantages of Hybrid Topology

- Complex in design.
- Costly.

NETWORK SWITCH

A network switch is a hardware device that channels incoming data from multiple input ports to a specific output port that will take it toward its intended destination. It is a small device that transfers data packets between multiple network devices such as computers, routers, servers or other switches.

In a local area network (LAN) using Ethernet, a network switch determines where to send each incoming message frame by looking at the physical device address (also known as the Media Access Control address or MAC address). Switches maintain tables that match each MAC address to the port which the MAC address is received.

A network switch operates on the network layer, called layer 2 of the OSI model.

Network Device Layers

Network devices can be separated by the layer they operate on, defined by the OSI model. The OSI model conceptualizes networks separating protocols by layers. Control is typically passed from one layer to the next. Some layers include:

- Layer 1- or the physical layer or below, which can transfer data but cannot manage the traffic coming through it. An example would be Ethernet hubs or cables.
- Layer 2- or the data link layer, which uses hardware addresses to receive and forward data. A network switch is an example of what type of device is on layer 2.
- Layer 3- or the network layer, which performs similar functions to a router and also supports multiple kinds of physical networks on different ports. Examples include routers or layer 3 switches.

Other layers include layer 4 (the transport layer), layer 5 (the session layer), layer 6 (the presentation layer) and layer 7 (the application layer).

Working of a Network Switch

Fundamental Concepts of a Networking Switch

Switches, physical and virtual, comprise the vast majority of network devices in modern data networks. They provide the wired connections to desktop computers, wireless access points, industrial machinery and some internet of things (IoT) devices such as card entry systems. They interconnect the computers that host virtual machines (VMs) in data centers, as well as the dedicated physical servers, and much of the storage infrastructure. They carry vast amounts of traffic in telecommunications provider networks.

A network switch can be deployed in the following ways:

- Edge, or access switches: These switches manage traffic either coming into or exiting the network. Devices like computers and access points connect to edge switches.
- Aggregation, or distribution switches: These switches are placed within an optional middle layer. Edge switches connect into these and they can send traffic from switch to switch or send it up to core switches.
- Core switches: These network switches comprise the backbone of the network, connecting either aggregation or edge switches, connecting user or device edge networks to data center networks and, typically, connecting enterprise LANs to the routers that connect them to the internet.

If a frame is forwarded to a MAC address unknown to the switch infrastructure, it is flooded to all ports in the switching domain. Broadcast and multicast frames are also flooded. This is known as BUM flooding - broadcast, unknown unicast, and multicast flooding. This capability makes a switch a Layer 2 or data-link layer device in the Open Systems Interconnection (OSI) communications model.

Many data centers adopt a leaf/spine architecture, which eliminates the aggregation layer. In this design, servers and storage connect to leaf switches (edge switches) and every leaf switch connects into two or more spine (core) switches. This minimizes the number of hops data has to take getting from source to destination, and, thereby, reduces the time spent in transit, or latency.

Some data centers establish a fabric or mesh network design that makes every device appear to be on a single, large switch. This approach reduces latency to its minimum and is used for highly demanding applications such as high-performance computing (HPC) in financial services or engineering.

Not all networks use switches. For example, a network may be (and often was, in the 1980s and 1990s) organized in a token ring or connected via a bus or a hub or repeater. In these networks, every connected device sees all traffic and reads the traffic addressed to it. A network can also be established by directly connecting computers to one another, without a separate layer of network devices; this approach is mostly of interest in HPC contexts where sub-5-microsecond latencies are desired and can become quite complex to design, wire and manage.

Types of Networking Switches

There are several types of switches in networking in addition to physical devices:

- Virtual switches are software-only switches instantiated inside VM hosting environments.
- A routing switch connects LANs; in addition to doing MAC-based Layer 2 switching it can also perform routing functions at OSI Layer 3 (the network layer) directing traffic based on the Internet Protocol (IP) address in each packet.

- A managed switch which lets a user adjust each port on the switch, allowing monitoring and configuration changes.
- An unmanaged switch which allows Ethernet devices to pass data automatically utilizing auto-negotiation (which determines parameters such as the data rate). The configuration is fixed and cannot be edited.
- Smart Switches, also called partially managed switches, which can be configured to allow more control over data transmissions but have more limitations compared to managed switches.

Network Switches vs. Hubs and Routers

Network switches can be similar looking to both hubs and routers; however, they have different functionalities and operate on separate layers. For example, a hub is relatively simple compared to a network switch. The goal of a hub is to connect all the nodes in a network. Because a hub can't manage data going in and out of it as a network switch can, there are a lot of communication collisions. Hubs are a layer 1 physical device, compared to a network switch which is a layer 2 on the OSI model.

A router is a device which joins networks and routes traffic between them. Routers are a layer 3 device on the OSI model and will deal with IP addresses. IP addresses route packets across the internet. As an example, an individual's router will connect their local network to their ISPs network.

TELECOMMUNICATIONS LINK

In telecommunications a link is a communication channel that connects two or more devices. The link may be physical or logical that uses one or more physical links or shares a physical link with other telecommunications links.

A telecommunications link is generally one of several types of information transmission paths such as those provided by communication satellites, terrestrial radio communications infrastructure and computer networks to connect two or more points.

The term *link* is widely used in computer networking to refer to the communications facilities that connect nodes of a network. When the link is a logical link the type of physical link should always be specified (e.g., data link, uplink, downlink, fiber optic link, point-to-point link, etc.)

Types

Point-to-Point

A point-to-point link is a dedicated link that connects exactly two communication facilities (e.g., two nodes of a network, an intercom station at an entryway with a single internal intercom station, a radio path between two points, etc.).

Broadcast

Broadcast links connect two or more nodes and support *broadcast transmission*, where one node can transmit so that all other nodes can receive the same transmission. Ethernet is an example.

Multipoint

Also known as a *multidrop* link, a multipoint link is a link that connects *two or more* nodes. Also known as general topology networks, these include ATM and Frame Relay links, as well as X.25 networks when used as links for a network layer protocol like IP. Unlike broadcast links, there is no mechanism to efficiently send a single message to all other nodes without copying and retransmitting the message.

Point-to-Multipoint

A point-to-multipoint link (or simply a *multipoint*) is a specific type of multipoint link which consists of a central connection endpoint (CE) that is connected to multiple peripheral CEs. Any transmission of data that originates from the central CE is received by all of the peripheral CEs while any transmission of data that originates from any of the peripheral CEs is only received by the central CE.

Private and Public — Accessibility and Ownership

Links are often referred to by terms which refer to the ownership and accessibility of the link.

- A private link is a link that is either owned by a specific entity or a link that is only accessible by a specific entity.
- A public link is a link that uses the public switched telephone network or other public utility or entity to provide the link and which may also be accessible by anyone.

Direction

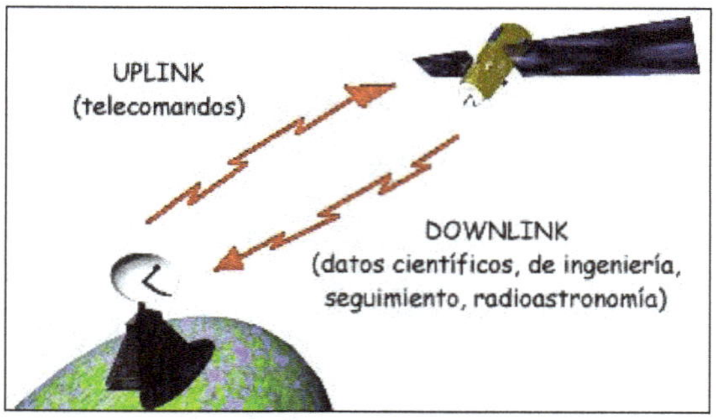

Feeder links, here: uplink/downlink.

Uplink

- Pertaining to radiocommunication service, an uplink (UL or U/L) is the portion of a feeder link used for the transmission of signals from an earth station to a space radio station, space radio system or high altitude platform station.

- Pertaining to GSM and cellular networks, the radio uplink is the transmission path from the mobile station (cell phone) to a base station (cell site). Traffic and signalling flowing within the BSS and NSS may also be identified as uplink and downlink.
- Pertaining to computer networks, an uplink is a connection from data communications equipment toward the network core. This is also known as an upstream connection.

Downlink

- Pertaining to radiocommunication service, a downlink (DL or D/L) is the portion of a feeder link used for the transmission of signals from a space radio station, space radio system or high altitude platform station to an earth station.
- In the context of satellite communications, a downlink (DL) is the link from a satellite to a ground station.
- Pertaining to cellular networks, the radio downlink is the transmission path from a cell site to the cell phone. Traffic and signalling flowing within the base station subsystem (BSS) and network switching subsystem(NSS) may also be identified as uplink and downlink.
- Pertaining to computer networks, a downlink is a connection from data communications equipment towards data terminal equipment. This is also known as a downstream connection.

Forward Link

A forward link is the link from a fixed location (e.g., a base station) to a mobile user. If the link includes a communications relay satellite, the forward link will consist of both an uplink (base station to satellite) and a downlink (satellite to mobile user).

Reverse Link

The reverse link (sometimes called a *return channel*) is the link from a mobile user to a fixed base station.

If the link includes a communications relay satellite, the reverse link will consist of both an uplink (mobile station to satellite) and a downlink (satellite to base station) which together constitute a half hop.

NODE NETWORKING

In telecommunications networks, a node is either a redistribution point or a communication endpoint. The definition of a node depends on the network and protocol layer referred to. A physical network node is an active electronic device that is attached to a network, and is capable of creating, receiving, or transmitting information over a communications channel. A passive distribution point such as a distribution frame or patch panel is consequently not a node.

In data communication, a physical network node may either be data communication equipment (DCE) such as a modem, hub, bridge or switch; or data terminal equipment (DTE) such as a digital telephone handset, a printer or a host computer.

If the network in question is a local area network (LAN) or wide area network (WAN), every LAN or WAN node, that are at least data link layer devices, must have a network address, typically one for each network interface controller it possesses. Examples are computers, packet switches, xDSL modems (with Ethernet interface) and wireless LAN access points. Equipment, such as a hub, repeater or PSTN modem with serial interface, that operate only below the data link layer does not require a network address.

If the network in question is the Internet or an Intranet, many physical network nodes are host computers, also known as Internet nodes, identified by an IP address, and all hosts are physical network nodes. However, some data link layer devices such as switches, bridges and wireless access points do not have an IP host address (except sometimes for administrative purposes), and are not considered to be Internet nodes or hosts, but as physical network nodes and LAN nodes.

In the fixed telephone network, a node may be a public or private telephone exchange, a remote concentrator or a computer providing some intelligent network service. In cellular communication, switching points and databases such as the Base station controller, Home Location Register, Gateway GPRS Support Node (GGSN) and Serving GPRS Support Node (SGSN) are examples of nodes. Cellular network base stations are not considered to be nodes in this context.

In cable television systems (CATV), this term has assumed a broader context and is generally associated with a fiber optic node. This can be defined as those homes or businesses within a specific geographic area that are served from a common fiber optic receiver. A fiber optic node is generally described in terms of the number of "homes passed" that are served by that specific fiber node.

Distributed Systems

If the network in question is a distributed system, the nodes are clients, servers or peers. A peer may sometimes serve as client, sometimes server. In a peer-to-peer or overlay network, nodes that actively route data for the other networked devices as well as themselves are called supernodes.

Distributed systems may sometimes use *virtual nodes* so that the system is not oblivious to the heterogeneity of the nodes. This issue is addressed with special algorithms, like consistent hashing, as it is the case in Amazon's Dynamo.

Cloud Computing

Within a vast computer network, the individual computers on the periphery of the network, those that do not also connect other networks, and those that often connect transiently to one or more clouds are called end nodes. Typically, within the cloud computing construct, the individual user/customer computer that connects into one well-managed cloud is called an end node. Since these computers are a part of the network yet unmanaged by the cloud's host, they present significant risks to the entire cloud. This is called the end node problem. There are several means to remedy this problem but all require instilling trust in the end node computer.

PACKET SWITCHING

Packet switching is a method of grouping data that is transmitted over a digital network into *packets*. Packets are made of a header and a payload. Data in the header are used by networking hardware to direct the packet to its destination where the payload is extracted and used by application software. Packet switching is the primary basis for data communications in computer networks worldwide.

In the early 1960s, American computer scientist Paul Baran developed the concept *Distributed Adaptive Message Block Switching* with the goal to provide a fault-tolerant, efficient routing method for telecommunication messages as part of a research program at the RAND Corporation, funded by the US Department of Defense. This concept contrasted with, and contradicted, then-established principles of pre-allocation of network bandwidth, largely fortified by the development of telecommunications in the Bell System. The new concept found little resonance among network implementers until the independent work of British computer scientist Donald Davies at the National Physical Laboratory (United Kingdom) in 1965. Davies is credited with coining the modern term *packet switching* and inspiring numerous packet switching networks in the decade following, including the incorporation of the concept in the early ARPANET in the United States.

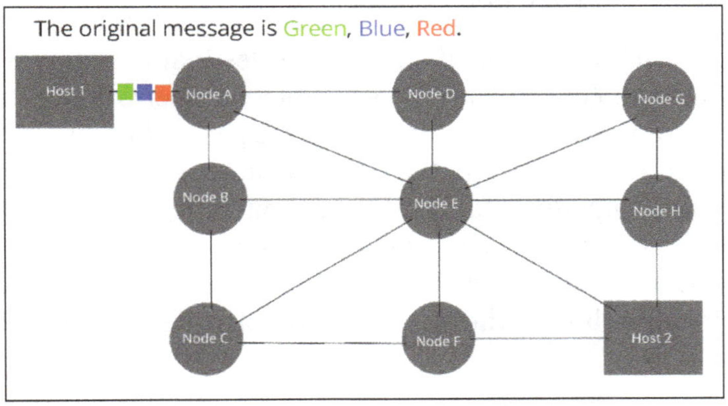

An animation demonstrating datagram type of packet switching across a network.

A simple definition of packet switching is: The routing and transferring of data by means of addressed packets so that a channel is occupied during the transmission of the packet only, and upon completion of the transmission the channel is made available for the transfer of other traffic.

Packet switching allows delivery of variable bit rate data streams, realized as sequences of packets, over a computer network which allocates transmission resources as needed using statistical multiplexing or dynamic bandwidth allocation techniques. As they traverse networking hardware, such as switches and routers, packets are received, buffered, queued, and retransmitted (stored and forwarded), resulting in variable latency and throughput depending on the link capacity and the traffic load on the network. Packets are normally forwarded by intermediate network nodes asynchronously using first-in, first-out buffering, but may be forwarded according to some scheduling discipline for fair queuing, traffic shaping, or for differentiated or guaranteed quality of service, such as weighted fair queuing or leaky bucket. Packet-based communication may be implemented with or without intermediate forwarding nodes (switches and routers). In case of a shared physical medium (such as radio or 10BASE5), the packets may be delivered according to a multiple access scheme.

Packet switching contrasts with another principal networking paradigm, circuit switching, a method which pre-allocates dedicated network bandwidth specifically for each communication session, each having a constant bit rate and latency between nodes. In cases of billable services, such as cellular communication services, circuit switching is characterized by a fee per unit of connection time, even when no data is transferred, while packet switching may be characterized by a fee per unit of information transmitted, such as characters, packets, or messages.

A packet switch has four components: input ports, output ports, routing processor, and switching fabric.

Connectionless and Connection-Oriented Modes

Packet switching may be classified into connectionless packet switching, also known as datagram switching, and connection-oriented packet switching, also known as virtual circuit switching. Examples of connectionless systems are Ethernet, Internet Protocol (IP), and the User Datagram Protocol (UDP). Connection-oriented systems include X.25, Frame Relay, Multiprotocol Label Switching (MPLS), and the Transmission Control Protocol (TCP).

In connectionless mode each packet is labeled with a destination address, source address, and port numbers. It may also be labeled with the sequence number of the packet. This information eliminates the need for a pre-established path to help the packet find its way to its destination, but means that more information is needed in the packet header, which is therefore larger. The packets are routed individually, sometimes taking different paths resulting in out-of-order delivery. At the destination, the original message may be reassembled in the correct order, based on the packet sequence numbers. Thus a virtual circuit carrying a byte stream is provided to the application by a transport layer protocol, although the network only provides a connectionless network layer service.

Connection-oriented transmission requires a setup phase to establish the parameters of communication before any packet is transferred. The signaling protocols used for setup allow the application to specify its requirements and discover link parameters. Acceptable values for service parameters may be negotiated. The packets transferred may include a connection identifier rather than address information and the packet header can be smaller, as it only needs to contain this code and any information, such as length, timestamp, or sequence number, which is different for different packets. In this case, address information is only transferred to each node during the connection setup phase, when the route to the destination is discovered and an entry is added to the switching table in each network node through which the connection passes. When a connection identifier is used, routing a packet requires the node to look up the connection identifier in a table.

Connection-oriented transport layer protocols such as TCP provide a connection-oriented service by using an underlying connectionless network. In this case, the end-to-end principle dictates that the end nodes, not the network itself, are responsible for the connection-oriented behavior.

Packet Switching in Networks

Packet switching is used to optimize the use of the channel capacity available in digital telecommunication networks, such as computer networks, and minimize the transmission latency (the time it takes for data to pass across the network), and to increase robustness of communication.

Packet switching is used in the Internet and most local area networks. The Internet is implemented by the Internet Protocol Suite using a variety of Link Layer technologies. For example, Ethernet and Frame Relay are common. Newer mobile phone technologies (e.g., GPRS, i-mode) also use packet switching. Packet switching is associated with connectionless networking because, in these systems, no connection agreement needs to be established between communicating parties prior to exchanging data.

X.25 is a notable use of packet switching in that, despite being based on packet switching methods, it provides virtual circuits to the user. These virtual circuits carry variable-length packets. In 1978, X.25 provided the first international and commercial packet switching network, the International Packet Switched Service (IPSS). Asynchronous Transfer Mode (ATM) also is a virtual circuit technology, which uses fixed-length cell relay connection oriented packet switching. Technologies such as Multiprotocol Label Switching (MPLS) and the Resource Reservation Protocol (RSVP) create virtual circuits on top of datagram networks. MPLS and its predecessors, as well as ATM, have been called "fast packet" technologies. MPLS, indeed, has been called "ATM without cells". Virtual circuits are especially useful in building robust failover mechanisms and allocating bandwidth for delay-sensitive applications.

Packet-switched Networks

The history of packet-switched networks can be divided into three overlapping eras: early networks before the introduction of X.25 and the OSI model, the X.25 era when many postal, telephone, and telegraph companies introduced networks with X.25 interfaces, and the Internet era.

Early Networks

Research into packet switching at the National Physical Laboratory (NPL) began with a proposal for a wide-area network in 1965, and a local-area network in 1966. ARPANET funding was secured in 1966 by Bob Taylor, and planning began in 1967 when he hired Larry Roberts. The NPL network, ARPANET, and SITA HLN became operational in 1969. Before the introduction of X.25 in 1973, about twenty different network technologies had been developed. Two fundamental differences involved the division of functions and tasks between the hosts at the edge of the network and the network core. In the datagram system, the hosts have the responsibility to ensure orderly delivery of packets. The User Datagram Protocol (UDP) is an example of a datagram protocol. In the virtual call system, the network guarantees sequenced delivery of data to the host. This results in a simpler host interface with less functionality than in the datagram model. The X.25 protocol suite uses this network type.

AppleTalk

AppleTalk is a proprietary suite of networking protocols developed by Apple in 1985 for Apple Macintosh computers. It was the primary protocol used by Apple devices through the 1980s and

1990s. AppleTalk included features that allowed local area networks to be established *ad hoc* without the requirement for a centralized router or server. The AppleTalk system automatically assigned addresses, updated the distributed namespace, and configured any required inter-network routing. It was a plug-n-play system.

AppleTalk versions were also released for the IBM PC and compatibles, and the Apple IIGS. AppleTalk support was available in most networked printers, especially laser printers, some file servers and routers. AppleTalk support was terminated in 2009, replaced by TCP/IP protocols.

ARPANET

The ARPANET was a progenitor network of the Internet and the first network to run the TCP/IP suite using packet switching technologies.

BNRNET

BNRNET was a network which Bell Northern Research developed for internal use. It initially had only one host but was designed to support many hosts. BNR later made major contributions to the CCITT X.25 project.

CYCLADES

The CYCLADES packet switching network was a French research network designed and directed by Louis Pouzin. First demonstrated in 1973, it was developed to explore alternatives to the early Arpanet design and to support network research generally. It was the first network to make the hosts responsible for reliable delivery of data, rather than the network itself, using unreliable datagrams and associated end-to-end protocol mechanisms. Concepts of this network influenced later Arpanet architecture.

DECnet

DECnet is a suite of network protocols created by Digital Equipment Corporation, originally released in 1975 in order to connect two PDP-11 minicomputers. It evolved into one of the first peer-to-peer network architectures, thus transforming DEC into a networking powerhouse in the 1980s. Initially built with three layers, it later (1982) evolved into a seven-layer OSI-compliant networking protocol. The DECnet protocols were designed entirely by Digital Equipment Corporation. However, DECnet Phase II (and later) were open standards with published specifications, and several implementations were developed outside DEC, including one for Linux.

DDX-1

This was an experimental network from Nippon PTT. It mixed circuit switching and packet switching. It was succeeded by DDX-2.

EIN Née COST II

European Informatics Network was a project to link several national networks. It became operational in 1976.

EPSS

The Experimental Packet Switching System (EPSS) was an experiment of the UK Post Office. It was the first public packet switching network in the UK when it began operating in 1977, based on protocols defined by the UK academic community in 1975. Ferranti supplied the hardware and software. The handling of link control messages (acknowledgements and flow control) was different from that of most other networks.

GEIS

As General Electric Information Services (GEIS), General Electric was a major international provider of information services. The company originally designed a telephone network to serve as its internal (albeit continent-wide) voice telephone network.

In 1965, at the instigation of Warner Sinback, a data network based on this voice-phone network was designed to connect GE's four computer sales and service centers (Schenectady, New York, Chicago, and Phoenix) to facilitate a computer time-sharing service, apparently the world's first commercial online service. (In addition to selling GE computers, the centers were computer service bureaus, offering batch processing services. They lost money from the beginning, and Sinback, a high-level marketing manager, was given the job of turning the business around. He decided that a time-sharing system, based on Kemeny's work at Dartmouth—which used a computer on loan from GE—could be profitable. Warner was right.)

After going international some years later, GEIS created a network data center near Cleveland, Ohio. Very little has been published about the internal details of their network. (Though it has been stated by some that Tymshare copied the GEIS system to create their network, Tymnet.) The design was hierarchical with redundant communication links.

IPSANET

IPSANET was a semi-private network constructed by I. P. Sharp Associates to serve their time-sharing customers. It became operational in May 1976.

IPX/SPX

The Internetwork Packet Exchange (IPX) and Sequenced Packet Exchange (SPX) are Novell networking protocols derived from Xerox Network Systems' IDP and SPP protocols, respectively. They were used primarily on networks using the Novell NetWare operating systems.

Merit Network

Merit Network, Inc., an independent non-profit 501(c)(3) corporation governed by Michigan's public universities, was formed in 1966 as the Michigan Educational Research Information Triad to explore computer networking between three of Michigan's public universities as a means to help the state's educational and economic development. With initial support from the State of Michigan and the National Science Foundation (NSF), the packet-switched network was first demonstrated in December 1971 when an interactive host-to-host connection was made between the IBM mainframe computer systems at the University of Michigan in Ann Arbor and Wayne State University in Detroit. In October 1972, connections to the CDC mainframe at Michigan State University in East Lansing completed the triad. Over the next several years, in addition to host-to-host interactive connections, the network was enhanced to support terminal-to-host connections, host-to-host batch connections (remote job submission, remote printing, batch file transfer), interactive file transfer, gateways to the Tymnet and Telenet public data networks, X.25 host attachments, gateways to X.25 data networks, Ethernet attached hosts, and eventually TCP/IP; additionally, public universities in Michigan joined the network. All of this set the stage for Merit's role in the NSFNET project starting in the mid-1980s.

NPL

In 1965, Donald Davies of the National Physical Laboratory (United Kingdom) designed and proposed a national data network based on packet switching. The proposal was not taken up nationally, but by 1967, a pilot experiment had demonstrated the feasibility of packet switched networks.

By 1969 Davies had begun building the Mark I packet-switched network to meet the needs of the multidisciplinary laboratory and prove the technology under operational conditions. In 1976, 12 computers and 75 terminal devices were attached, and more were added until the network was replaced in 1986. NPL, followed by ARPANET, were the first two networks in the world to use packet switching, and were interconnected in the early 1970s.

OCTOPUS

Octopus was a local network at Lawrence Livermore National Laboratory. It connected sundry hosts at the lab to interactive terminals and various computer peripherals including a bulk storage system.

Philips Research

Philips Research Laboratories in Redhill, Surrey developed a packet switching network for internal use. It was a datagram network with a single switching node.

PUP

PARC Universal Packet (PUP or Pup) was one of the two earliest internetwork protocol suites; it was created by researchers at Xerox PARC in the mid-1970s. The entire suite provided routing and packet delivery, as well as higher level functions such as a reliable byte stream, along with numerous applications. Further developments led to Xerox Network Systems (XNS).

RCP

RCP was an experimental network created by the French PTT. It was used to gain experience with packet switching technology before the specification of TRANSPAC was frozen. RCP was a virtual-circuit network in contrast to CYCLADES which was based on datagrams. RCP emphasised terminal-to-host and terminal-to-terminal connection; CYCLADES was concerned with host-to-host communication. TRANSPAC was introduced as an X.25 network. RCP influenced the specification of X.25

RETD

Red Especial de Transmisión de Datos was a network developed by Compañía Telefónica Nacional de España. It became operational in 1972 and thus was the first public network.

SCANNET

"The experimental packet-switched Nordic telecommunication network SCANNET was implemented in Nordic technical libraries in the 1970s, and it included first Nordic electronic journal Extemplo. Libraries were also among first ones in universities to accommodate microcomputers for public use in the early 1980s."

SITA HLN

SITA is a consortium of airlines. Its High Level Network became operational in 1969 at about the same time as ARPANET. It carried interactive traffic and message-switching traffic. As with many non-academic networks, very little has been published about it.

IBM Systems Network Architecture

IBM Systems Network Architecture (SNA) is IBM's proprietary networking architecture created in 1974. An IBM customer could acquire hardware and software from IBM and lease private lines from a common carrier to construct a private network.

Telenet

Telenet was the first FCC-licensed public data network in the United States. It was founded by former ARPA IPTO director Larry Roberts as a means of making ARPANET technology public. He had tried to interest AT&T in buying the technology, but the monopoly's reaction was that this was incompatible with their future. Bolt, Beranack and Newman (BBN) provided the financing. It initially used ARPANET technology but changed the host interface to X.25 and the terminal interface to X.29. Telenet designed these protocols and helped standardize them in the CCITT. Telenet was incorporated in 1973 and started operations in 1975. It went public in 1979 and was then sold to GTE.

Tymnet

Tymnet was an international data communications network headquartered in San Jose, CA that utilized virtual call packet switched technology and used X.25, SNA/SDLC, BSC and ASCII interfaces to connect host computers (servers) at thousands of large companies, educational institutions, and government agencies. Users typically connected via dial-up connections or dedicated async connections. The business consisted of a large public network that supported dial-up users and a private network business that allowed government agencies and large companies (mostly banks and airlines) to build their own dedicated networks. The private networks were often connected via gateways to the public network to reach locations not on the private network. Tymnet was also connected to dozens of other public networks in the U.S. and internationally via X.25/X.75 gateways.

XNS

Xerox Network Systems (XNS) was a protocol suite promulgated by Xerox, which provided routing and packet delivery, as well as higher level functions such as a reliable stream, and remote procedure calls. It was developed from PARC Universal Packet (PUP).

X.25 Era

There were two kinds of X.25 networks. Some such as DATAPAC and TRANSPAC were initially implemented with an X.25 external interface. Some older networks such as TELENET and TYMNET were modified to provide a X.25 host interface in addition to older host connection schemes. DATAPAC was developed by Bell Northern Research which was a joint venture of Bell Canada (a common carrier) and Northern Telecom (a telecommunications equipment supplier). Northern Telecom sold several DATAPAC clones to foreign PTTs including the Deutsche Bundespost. X.75 and X.121 allowed the interconnection of national X.25 networks. A user or host could call a host on a foreign network by including the DNIC of the remote network as part of the destination address.

AUSTPAC

AUSTPAC was an Australian public X.25 network operated by Telstra. Started by Telecom Australia in the early 1980s, AUSTPAC was Australia's first public packet-switched data network, supporting applications such as on-line betting, financial applications—the Australian Tax Office made use of AUSTPAC—and remote terminal access to academic institutions, who maintained their connections to AUSTPAC up until the mid-late 1990s in some cases. Access can be via a dial-up terminal to a PAD, or, by linking a permanent X.25 node to the network.

ConnNet

ConnNet was a packet-switched data network operated by the Southern New England Telephone Company serving the state of Connecticut.

Datanet 1

Datanet 1 was the public switched data network operated by the Dutch PTT Telecom (now known as KPN). Strictly speaking Datanet 1 only referred to the network and the connected users via leased lines, the name also referred to the public PAD service *Telepad*. And because the main Videotex service used the network and modified PAD devices as infrastructure the name Datanet 1 was used for these services as well. Although this use of the name was incorrect all these services were managed by the same people within one department of KPN contributed to the confusion.

Datapac

DATAPAC was the first operational X.25 network. It covered major Canadian cities and was eventually extended to smaller centres.

Datex-P

Deutsche Bundespost operated this national network in Germany. The technology was acquired from Northern Telecom.

Eirpac

Eirpac is the Irish public switched data network supporting X.25 and X.28. It was launched in 1984, replacing Euronet. Eirpac is run by Eircom.

HIPA-NET

Hitachi designed a private network system for sale as a turnkey package to multi-national organizations. In addition to providing X.25 packet switching, message switching software was also included. Messages were buffered at the nodes adjacent to the sending and receiving terminals. Switched virtual calls were not supported, but through the use of "logical ports" an originating terminal could have a menu of pre-defined destination terminals.

Iberpac

Iberpac is the Spanish public packet-switched network, providing X.25 services. Iberpac is run by Telefonica.

JANET

JANET was the UK academic and research network, linking all universities, higher education establishments, publicly funded research laboratories. The X.25 network was based mainly on GEC 4000 series switches, and run X.25 links at up to 8 Mbit/s in its final phase before being converted to an IP based network. The JANET network grew out of the 1970s SRCnet (later called SERCnet) network.

PSS

Packet Switch Stream (PSS) was the UK Post Office (later to become British Telecom) national X.25 network with a DNIC of 2342. British Telecom renamed PSS under its GNS (Global Network Service) name, but the PSS name has remained better known. PSS also included public dial-up PAD access, and various InterStream gateways to other services such as Telex.

TRANSPAC

TRANSPAC was the national X.25 network in France. It was developed locally at about the same time as DATAPAC in Canada. The development was done by the French PTT and influenced by the experimental RCP network. It began operation in 1978, and served both commercial users and, after Minitel began, consumers.

VENUS-P

VENUS-P was an international X.25 network that operated from April 1982 through March 2006. At its subscription peak in 1999, VENUS-P connected 207 networks in 87 countries.

Venepaq

Venepaq is the national X.25 public network in Venezuela. It is run by Cantv and allow direct connection and dial up connections. Provides nationalwide access at very low cost. It provides national and international access. Venepaq allow connection from 19.2 kbit/s to 64 kbit/s in direct connections, and 1200, 2400 and 9600 bit/s in dial up connections.

Internet Era

When Internet connectivity was made available to anyone who could pay for an ISP subscription, the distinctions between national networks blurred. The user no longer saw network identifiers such as the DNIC. Some older technologies such as circuit switching have resurfaced with new names such as fast packet switching. Researchers have created some experimental networks to complement the existing Internet.

CSNET

The Computer Science Network (CSNET) was a computer network funded by the U.S. National Science Foundation (NSF) that began operation in 1981. Its purpose was to extend networking benefits, for computer science departments at academic and research institutions that could not be directly connected to ARPANET, due to funding or authorization limitations. It played a significant role in spreading awareness of, and access to, national networking and was a major milestone on the path to development of the global Internet.

Internet2

Internet2 is a not-for-profit United States computer networking consortium led by members from the research and education communities, industry, and government. The Internet2 community, in partnership with Qwest, built the first Internet2 Network, called Abilene, in 1998 and was a prime investor in the National LambdaRail (NLR) project. In 2006, Internet2 announced a partnership with Level 3 Communications to launch a brand new nationwide network, boosting its capacity from 10 Gbit/s to 100 Gbit/s. In October, 2007, Internet2 officially retired Abilene and now refers to its new, higher capacity network as the Internet2 Network.

NSFNET

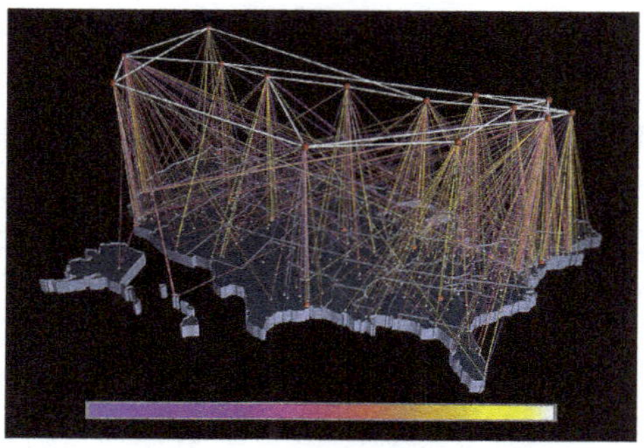

NSFNET Traffic 1991, NSFNET backbone nodes are shown at the top, regional networks below, traffic volume is depicted from purple (zero bytes) to white (100 billion bytes), visualization by NCSA using traffic data provided by the Merit Network.

The National Science Foundation Network (NSFNET) was a program of coordinated, evolving projects sponsored by the National Science Foundation (NSF) beginning in 1985 to promote advanced research and education networking in the United States. NSFNET was also the name given to several nationwide backbone networks operating at speeds of 56 kbit/s, 1.5 Mbit/s (T1), and 45 Mbit/s (T3) that were constructed to support NSF's networking initiatives from 1985-1995. Initially created to link researchers to the nation's NSF-funded supercomputing centers, through further public funding and private industry partnerships it developed into a major part of the Internet backbone.

National LambdaRail

The National LambdaRail was launched in September 2003. It is a 12,000-mile high-speed national computer network owned and operated by the U.S. research and education community that runs over fiber-optic lines. It was the first transcontinental 10 Gigabit Ethernet network. It operates with high aggregate capacity of up to 1.6 Tbit/s and a high 40 Gbit/s bitrate, with plans for 100 Gbit/s. The upgrade never took place and NLR ceased operations in March 2014.

TransPAC, TransPAC2 and TransPAC3

TransPAC2 and TransPAC3, continuations of the TransPAC project, a high-speed international Internet service connecting research and education networks in the Asia-Pacific region to those in the US. TransPAC is part of the NSF's International Research Network Connections (IRNC) program.

Very High-speed Backbone Network Service

The Very high-speed Backbone Network Service (vBNS) came on line in April 1995 as part of a National Science Foundation (NSF) sponsored project to provide high-speed interconnection between NSF-sponsored supercomputing centers and select access points in the United States. The network was engineered and operated by MCI Telecommunications under a cooperative agreement with the NSF. By 1998, the vBNS had grown to connect more than 100 universities and research and engineering institutions via 12 national points of presence with DS-3 (45 Mbit/s), OC-3c (155 Mbit/s), and OC-12c (622 Mbit/s) links on an all OC-12c backbone, a substantial engineering feat for that time. The vBNS installed one of the first ever production OC-48c (2.5 Gbit/s) IP links in February 1999 and went on to upgrade the entire backbone to OC-48c.

In June 1999 MCI WorldCom introduced vBNS+ which allowed attachments to the vBNS network by organizations that were not approved by or receiving support from NSF. After the expiration of the NSF agreement, the vBNS largely transitioned to providing service to the government. Most universities and research centers migrated to the Internet2 educational backbone. In January 2006, when MCI and Verizon merged, vBNS+ became a service of Verizon Business.

TELEPHONE EXCHANGE

A telephone exchange is a telecommunications system used in the public switched telephone network or in large enterprises. An exchange consists of electronic components and in older systems also human operators that interconnect (*switch*) telephone subscriber lines or virtual circuits of digital systems to establish telephone calls between subscribers.

In historical perspective, telecommunication terms have been used with different semantics over time. The term telephone exchange is often used synonymously with central office (CO), a Bell System term. Often, a central office is defined as a building used to house the inside plant equipment of potentially several telephone exchanges, each serving a certain geographical area. Such an area has also been referred to as the exchange. Central office locations may also be identified in North America as wire centers, designating a facility from which a telephone obtains dial tone. For business and billing purposes, telephony carriers also define rate centers, which in larger cities may be clusters of central offices, to define specified geographical locations for determining distance measurements.

In the United States and Canada, the Bell System established in the 1940s a uniform system of identifying central offices with a three-digit central office code, that was used as a prefix to subscriber telephone numbers. All central offices within a larger region, typically aggregated by state, were assigned a common numbering plan area code. With the development of international and transoceanic telephone trunks, especially driven by direct customer dialing, similar efforts of systematic organization of the telephone networks occurred in many countries in the mid-20th century.

For corporate or enterprise use, a private telephone exchange is often referred to as a private branch exchange (PBX), when it has connections to the public switched telephone network. A PBX is installed in enterprise facilities, typically collocated with large office spaces or within an organizational campus to serve the local private telephone system and any private leased line circuits. Smaller installations might deploy a PBX or key telephone system in the office of a receptionist.

Technologies

Many terms used in telecommunication technology differ in meaning and usage among the various English speaking regions.

- *Manual service* is a condition in which a human telephone operator routes calls inside an exchange without the use of a dial.
- Dial service is when an exchange routes calls by a switch interpreting dialed digits.
- A telephone switch is the switching equipment of an exchange.
- A concentrator is a device that concentrates traffic, be it remote or co-located with the switch.
- An off-hook condition represents a circuit that is in use, e.g., when a phone call is in progress.
- An on-hook condition represents an idle circuit, i.e. no phone call is in progress.
- A wire center is the area served by a particular switch or central office.

Central office originally referred to switching equipment and its operators, it is also used generally for the building that houses switching and related inside plant equipment. In United States telecommunication jargon, a central office (C.O.) is a common carrier switching center Class 5 telephone switch in which trunks and local loops are terminated and switched. In the UK, a telephone exchange means an exchange building, and is also the name for a telephone switch.

Manual Service Exchanges

1924 PBX switchboard.

With manual service, the customer lifts the receiver off-hook and asks the operator to connect the call to a requested number. Provided that the number is in the same central office, and located on the operator's switchboard, the operator connects the call by plugging the ringing cord into the jack corresponding to the called customer's line. If the called party's line is on a different switchboard in the same office, or in a different central office, the operator plugs into the trunk for the destination switchboard or office and asks the operator answering (known as the "B" operator) to connect the call.

Most urban exchanges provided common-battery service, meaning that the central office provided power to the subscriber telephone circuits for operation of the transmitter, as well as for automatic signaling with rotary dials. In common-battery systems, the pair of wires from a subscriber's telephone to the exchange carry 48V (nominal) DC potential from the telephone company end across the conductors. The telephone presents an open circuit when it is on-hook or idle.

When a subscriber's phone is off-hook, it presents an electrical resistance across the line which causes current to flow through the telephone and wires to the central office. In a manually operated switchboard, this current flowed through a relay coil, and actuated a buzzer or a lamp on the operator's switchboard, signaling the operator to perform service.

In the largest cities, it took many years to convert every office to automatic equipment, such as a panel switch. During this transition period, once numbers were standardized to the 2L-4N or

2L-5N format (two-letter exchange name and either four or five digits), it was possible to dial a number located in a manual exchange and be connected without requesting operator assistance. The policy of the Bell System stated that customers in large cities should not need to be concerned with the type of office, whether they were calling a manual or an automatic office.

When a subscriber dialed the number of a manual station, an operator at the destination office answered the call after seeing the number on an indicator, and connected the call by plugging a cord into the outgoing circuit and ringing the destination station. For example, if a dial customer calling from TAylor 4725 dialed a number served by a manual exchange, e.g., ADams 1383-W, the call was completed, from the subscriber's perspective, exactly as a call to LEnnox 5813, in an automated exchange. The party line letters W, R, J, and M were only used in manual exchanges with jack-per-line party lines.

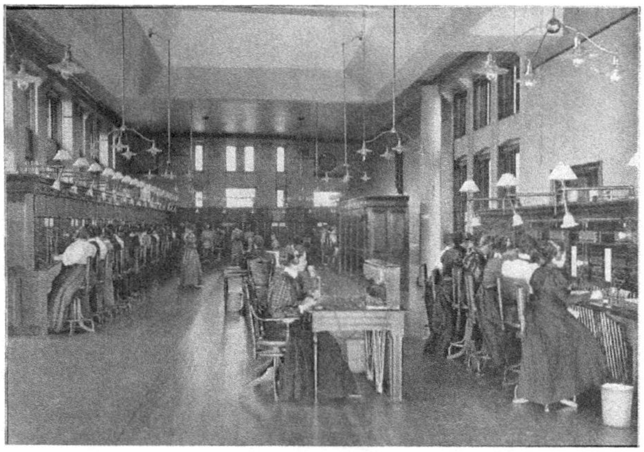

Montreal telephone exchange.

In contrast to the listing format Main *1234* for an automated office with two capital letters, a manual office, having listings such as Hillside 834 or East 23, was recognizable by the format in which the second letter was not capitalized.

Rural areas, as well as the smallest towns, had manual service and signaling was accomplished with magneto telephones, which had a crank for the signaling generator. To alert the operator, or another subscriber on the same line, the subscriber turned the crank to generate ringing current. The switchboard responded by interrupting the circuit, which dropped a metal tab above the subscriber's line jack and sounded a buzzer. Dry cell batteries, normally two large N°. 6 cells in the subscriber's telephone, provided the direct current for the transmitter. Such magneto systems were in use in the US as late as 1983, as in the small town, Bryant Pond, Woodstock, Maine.

Many small town magneto systems featured party lines, anywhere from two to ten or more subscribers sharing a single line. When calling a party, the operator used code ringing, a distinctive ringing signal sequence, such as two long rings followed by one short ring. Everyone on the line could hear the signals, and could pick up and monitor other people's conversations.

Early Automatic Exchanges

Automatic exchanges, or dial service, came into existence in the early 20th century. Their purpose was to eliminate the need for human switchboard operators who completed the connections required for a telephone call. Automation replaced human operators with electromechanical systems and telephones were equipped with a dial by which a caller transmitted the destination telephone number to the automatic switching system.

A rural telephone exchange building in Australia.

A telephone exchange automatically senses an off-hook condition of the telephone when the user removes the handset from the switchhook or cradle. The exchange provides dial tone at that time to indicate to the user that the exchange is ready to receive dialed digits. The pulses or DTMF tones generated by the telephone are processed and a connection is established to the destination telephone within the same exchange or to another distant exchange.

The exchange maintains the connection until one of the parties hangs up. This monitoring of connection status is called supervision. Additional features, such as billing equipment, may also be incorporated into the exchange.

The Bell System dial service implemented a feature called automatic number identification (ANI) which facilitated services like automated billing, toll-free 800-numbers, and 9-1-1 service. In manual service, the operator knows where a call is originating by the light on the switchboard jack field. Before ANI, long distance calls were placed into an operator queue and the operator asked the calling party's number and recorded it on a paper toll ticket.

Early exchanges were electromechanical systems using motors, shaft drives, rotating switches and relays. Some types of automatic exchanges were the Strowger switch or step-by-step switch, All Relay, X-Y, panel switch, Rotary system and the crossbar switch.

Electromechanical Signaling

Circuits interconnecting switches are called *trunks*. Before Signalling System 7, Bell System electromechanical switches in the United States originally communicated with one another over trunks using a variety of DC voltages and signaling tones, replaced today by digital signals.

Some signaling communicated dialed digits. An early form called Panel Call Indicator Pulsing used quaternary pulses to set up calls between a panel switch and a manual switchboard. Probably the most common form of communicating dialed digits between electromechanical switches was sending dial pulses, equivalent to a rotary dial's pulsing, but sent over trunk circuits between switches.

In Bell System trunks, it was common to use 20 pulse-per-second between crossbar switches and crossbar tandems. This was twice the rate of Western Electric/Bell System telephone dials. Using the faster pulsing rate made trunk utilization more efficient because the switch spent half as long listening to digits. DTMF was not used for trunk signaling.

Multi-frequency (MF) was the last of the pre-digital methods. It used a different set of tones sent in pairs like DTMF. Dialing was preceded by a special keypulse (KP) signal and followed by a start (ST). Variations of the Bell System MF tone scheme became a CCITT standard. Similar schemes were used in the Americas and in some European countries including Spain. Digit strings between switches were often abbreviated to further improve utilization.

For example, one switch might send only the last four or five digits of a telephone number. In one case, seven digit numbers were preceded by a digit 1 or 2 to differentiate between two area codes or office codes, (a two-digit-per-call savings). This improved revenue per trunk and reduced the number of digit receivers needed in a switch. Every task in electromechanical switches was done in big metallic pieces of hardware. Every fractional second cut off of call set up time meant fewer racks of equipment to handle call traffic.

Examples of signals communicating supervision or call progress include E and M signaling, SF signaling, and robbed-bit signaling. In physical (not carrier) E and M trunk circuits, trunks were four wire. Fifty trunks would require a hundred pair cable between switches, for example. Conductors in one common circuit configuration were named tip, ring, ear (E) and mouth (M). Tip and ring were the voice-carrying pair, and named after the tip and ring on the three conductor cords on the manual operator's console.

In two-way trunks with E and M signaling, a handshake took place to prevent both switches from colliding by dialing calls on the same trunk at the same time. By changing the state of these leads from ground to -48 volts, the switches stepped through a handshake protocol. Using DC voltage changes, the local switch would send a signal to get ready for a call and the remote switch would reply with an acknowledgment to go ahead with dial pulsing. This was done with relay logic and discrete electronics.

These voltage changes on the trunk circuit would cause pops or clicks that were audible to the subscriber as the electrical handshaking stepped through its protocol. Another handshake, to start timing for billing purposes, caused a second set of clunks when the called party answered.

A second common form of signaling for supervision was called single-frequency or *SF signaling*. The most common form of this used a steady 2,600 Hz tone to identify a trunk as idle. Trunk circuitry hearing a 2,600 Hz tone for a certain duration would go idle. (The duration requirement reduced falsing.) Some systems used tone frequencies over 3,000 Hz, particularly on SSB frequency division multiplex microwave radio relays.

On T-carrier digital transmission systems, bits within the T-1 data stream were used to transmit supervision. By careful design, the appropriated bits did not change voice quality appreciably. Robbed bits were translated to changes in contact states (opens and closures) by electronics in the channel bank hardware. This allowed direct current E and M signaling, or dial pulses, to be sent between electromechanical switches over a digital carrier which did not have DC continuity.

Sounds

A characteristic of electromechanical switching equipment is that the maintenance staff could hear the mechanical clattering of Strowgers, panel switches or crossbar relays. Most Bell System central offices were housed in reinforced concrete buildings with concrete ceilings and floors.

In rural areas some smaller switching facilities, such as community dial offices (CDOs), were housed in prefabricated metal buildings. These facilities almost always had concrete floors. The hard surfaces reflected sounds.

During heavy use periods, it could be difficult to converse in a central office switch room due to the clatter of calls being processed in a large switch. For example, on Mother's Day in the US, or on a Friday evening around 5pm, the metallic rattling could make raised voices necessary. For wire spring relay markers these noises resembled hail falling on a metallic roof.

On a pre-dawn Sunday morning, call processing might slow to the extent that one might be able to hear individual calls being dialed and set up. There were also noises from whining power inverters and whirring ringing generators. Some systems had a continual, rhythmic "clack-clack-clack" from wire spring relays that made reorder (120 ipm) and busy (60 ipm) signals.

Bell System installations typically had alarm bells, gongs, or chimes to announce alarms calling attention to a failed switch element. A trouble reporting card system was connected to switch common control elements. These trouble reporting systems punctured cardboard cards with a code that logged the nature of a failure. Reed relay technology in stored program control exchange finally quieted the environment.

Maintenance Tasks

A man operating a test board in an electromechanical switching office.

Electromechanical switching systems required sources of electricity in form of direct current (DC), as well as alternating ring current (AC), which were generated on-site with mechanical generators. In addition, telephone switches required adjustment of many mechanical parts. Unlike modern switches, a circuit connecting a dialed call through an electromechanical switch had DC continuity within the local exchange area via metallic conductors.

The design and maintenance procedures of all systems involved methods to avoid that subscribers experienced undue changes in the quality of the service or that they noticed failures. A variety of tools referred to as *make-busys* were plugged into electromechanical switch elements upon failure and during repairs. A make-busy identified the part being worked on as in-use, causing the switching logic to route around it. A similar tool was called a *TD tool.* Delinquent subscribers had their service temporarily denied (TDed). This was effected by plugging a tool into the subscriber's office equipment on Crossbar systems or line group in step-by-step switches. The subscriber could receive calls but could not dial out.

Strowger-based, step-by-step offices in the Bell System required continuous maintenance, such as cleaning. Indicator lights on equipment bays in step offices alerted staff to conditions such as blown fuses (usually white lamps) or a permanent signal (stuck off-hook condition, usually green indicators). Step offices were more susceptible to single-point failures than newer technologies.

Crossbar offices used more shared, common control circuits. For example, a digit receiver (part of an element called an Originating Register) would be connected to a call just long enough to collect the subscriber's dialed digits. Crossbar architecture was more flexible than step offices. Later crossbar systems had punch-card-based trouble reporting systems. By the 1970s, automatic number identification had been retrofitted to nearly all step-by-step and crossbar switches in the Bell System.

Electronic Switches

Electronic switching systems gradually evolved in stages from electromechanical hybrids with stored program control to the fully digital systems. Early systems used reed relay-switched metallic paths under digital control. Equipment testing, phone numbers reassignments, circuit lockouts and similar tasks were accomplished by data entry on a terminal.

Examples of these systems included the Western Electric 1ESS switch, Northern Telecom SP1, Ericsson AXE, Philips PRX/A, ITT Metaconta, British GPO/BT TXE series and several other designs were similar. Ericsson also developed a fully computerized version of their ARF crossbar exchange called ARE. These used a crossbar switching matrix with a fully computerized control system and provided a wide range of advanced services. Local versions were called ARE11 while tandem versions were known as ARE13. They were used in Scandinavia, Australia, Ireland and many other countries in the late 1970s and into the 1980s when they were replaced with digital technology.

These systems could use the old electromechanical signaling methods inherited from crossbar and step-by-step switches. They also introduced a new form of data communications: two 1ESS exchanges could communicate with one another using a data link called Common Channel Interoffice Signaling, (CCIS). This data link was based on CCITT 6, a predecessor to SS7. In European systems R2 signalling was normally used.

Digital Switches

Digital switches work by connecting two or more digital circuits, according to a dialed telephone number or other instruction. Calls are set up between switches. In modern networks, this is usually controlled using the Signalling System 7 (SS7) protocol, or one of its variants. Many networks around the world are now transitioning to voice over IP technologies which use Internet-based protocols such as the Session Initiation Protocol (SIP). These may have superseded TDM and SS7 based technologies in some networks.

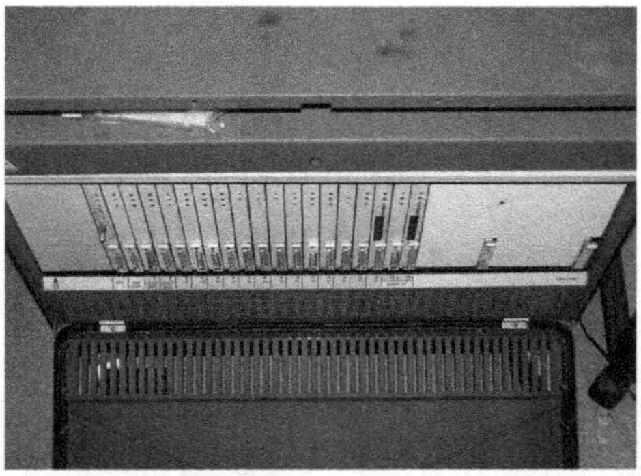

A typical satellite PABX with front cover removed.

The concepts of digital switching were developed by various labs in the United States and in Europe from the 1930s onwards. The first prototype digital switch was developed by Bell Labs as part of the ESSEX project while the first true digital exchange to be combined with digital transmission systems was designed by LCT (Laboratoire Central de Telecommunications) in Paris. The first digital switch to be placed into a public network was the Empress Exchange in London, England which was designed by the General Post Office research labs. This was a tandem switch that connected three Strowger exchanges in the London area. The first commercial roll-out of a fully digital local switching system was Alcatel's E10 system which began serving customers in Brittany in Northwestern France in 1972.

Prominent examples of digital switches include:

- Ericsson's AXE telephone exchange is the most widely used digital switching platform in the world and can be found throughout Europe and in most countries around the world. It is also very popular in mobile applications. This highly modular system was developed in Sweden in the 1970s as a replacement for the very popular range of Ericsson crossbar switches ARF, ARM, ARK and ARE used by many European networks from the 1950s onwards.

- Alcatel-Lucent inherited three of the world's most iconic digital switching systems : Alcatel E10, 1000-S12, and the Western Electric 5ESS.

  Alcatel developed the E10 system in France during the late 1960s and 1970s. This widely used family of digital switches was one of the earliest TDM switches to be widely used in public networks. Subscribers were first connected to E10A switches in France in 1972. This system is used in France, Ireland, China, and many other countries. It has been through many revisions and current versions are even integrated into All IP networks.

  Alcatel also acquired ITT System 12 which when it bought ITT's European operations. The S12 system and E10 systems were merged into a single platform in the 1990s. The S12 system is used in Germany, Italy, Australia, Belgium, China, India, and many other countries around the world.

  Finally, when Alcatel and Lucent merged, the company acquired Lucent's 5ESS and 4ESS systems used throughout the United States of America and in many other countries.

- Nokia Siemens Networks EWSD originally developed by Siemens, Bosch and DeTeWe for the German market is used throughout the world.

- Nortel now Genband DMS100 is very popular with operators all over the world.
- GTD-5 EAX developed by GTE Automatic Electric.
- NEC NEAX used in Japan, New Zealand and many other countries.
- Marconi System X originally developed by GPT and Plessey is a type of digital exchange used by BT Group in the UK public telephone network.

A digital exchange (Nortel DMS-100) used by an operator to offer local and long distance services in France. Each switch typically serves 10,000–100,000+ subscribers depending on the geographic area.

Digital switches encode the speech going on, in 8,000 time slices per second. At each time slice, a digital PCM representation of the tone is made. The digits are then sent to the receiving end of the line, where the reverse process occurs, to produce the sound for the receiving phone. In other words, when someone uses a telephone, the speaker's voice is "encoded" then reconstructed for the person on the other end. The speaker's voice is delayed in the process by a small fraction of one second — it is not "live", it is reconstructed — delayed only minutely.

Individual local loop telephone lines are connected to a remote concentrator. In many cases, the concentrator is co-located in the same building as the switch. The interface between remote concentrators and telephone switches has been standardised by ETSI as the V5 protocol. Concentrators are used because most telephones are idle most of the day, hence the traffic from hundreds or thousands of them may be concentrated into only tens or hundreds of shared connections.

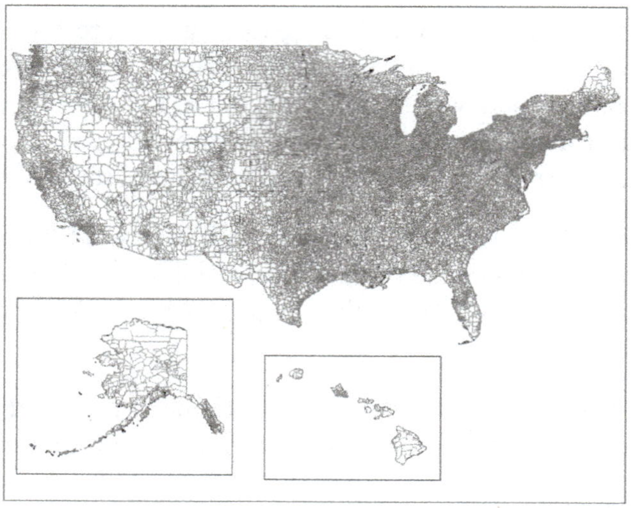

Map of the Wire Center locations in the US.

Some telephone switches do not have concentrators directly connected to them, but rather are used to connect calls between other telephone switches. These complex machines (or a series of them) in a central exchange building are referred to as "carrier-level" switches or tandem switches.

Some telephone exchange buildings in small towns now house only remote or satellite switches, and are homed upon a "parent" switch, usually several kilometres away. The remote switch is dependent on the parent switch for routing and number plan information. Unlike a digital loop carrier, a remote switch can route calls between local phones itself, without using trunks to the parent switch.

Telephone switches are usually owned and operated by a telephone service provider or *carrier* and located in their premises, but sometimes individual businesses or private commercial buildings will house their own switch, called a PBX, or Private branch exchange.

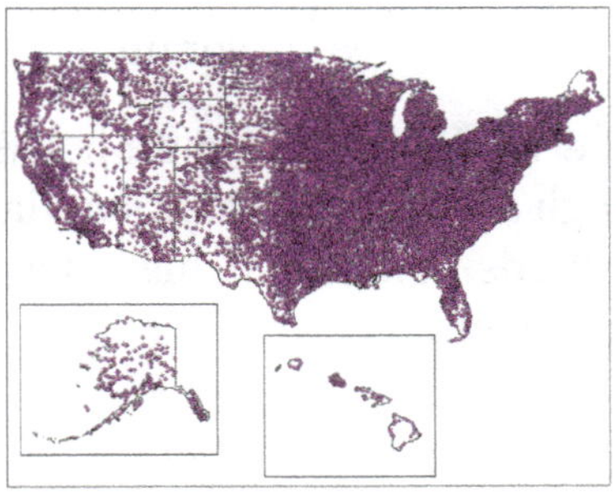

Map of the Central Office locations in the US.

Switch's Place in the System

Telephone switches are a small component of a large network. A major part, in terms of expense, maintenance, and logistics of the telephone system is outside plant, which is the wiring outside the central office. While many subscribers were served with party-lines in the middle of the 20th century, it was the goal that each subscriber telephone station were connected to an individual pair of wires from the switching system.

A typical central office may have tens of thousands of pairs of wires that appear on terminal blocks called the main distribution frame (MDF). A component of the MDF is protection: fuses or other devices that protect the switch from lightning, shorts with electric power lines, or other foreign voltages. In a typical telephone company, a large database tracks information about each subscriber pair and the status of each jumper. Before computerization of Bell System records in the 1980s, this information was handwritten in pencil in accounting ledger books.

To reduce the expense of outside plant, some companies use "pair gain" devices to provide telephone service to subscribers. These devices are used to provide service where existing copper facilities have been exhausted or by siting in a neighborhood, can reduce the length of copper pairs, enabling digital services such as Integrated Services Digital Network (ISDN) or digital subscriber line (DSL).

Pair gain or digital loop carriers (DLCs) are located outside the central office, usually in a large neighborhood distant from the CO. DLCs are often referred to as Subscriber Loop Carriers (SLCs), after a Lucent proprietary product.

DLCs can be configured as universal (UDLCs) or integrated (IDLCs). *Universal DLCs* have two terminals, a central office terminal (COT) and a remote terminal (RT), that function similarly. Both terminals interface with analog signals, convert to digital signals, and transport to the other side where the reverse is performed.

Sometimes, the transport is handled by separate equipment. In an *Integrated DLC*, the COT is eliminated. Instead, the RT is connected digitally to equipment in the telephone switch. This reduces the total amount of equipment required.

Switches are used in both local central offices and in long distance centers. There are two major types in the Public switched telephone network (PSTN), the Class 4 telephone switches designed for toll or switch-to-switch connections, and the Class 5 telephone switches or subscriber switches, which manage connections from subscriber telephones. Since the 1990s, hybrid Class 4/5 switching systems that serve both functions have become common.

Another element of the telephone network is time and timing. Switching, transmission and billing equipment may be slaved to very high accuracy 10 MHz standards which synchronize time events to very close intervals. Time-standards equipment may include Rubidium- or Caesium-based standards and a Global Positioning System receiver.

Switch Design

Long distance switches may use a slower, more efficient switch-allocation algorithm than local central offices, because they have near 100% utilization of their input and output channels. Central offices have more than 90% of their channel capacity unused.

Traditional telephone switches connected physical circuits (e.g., wire pairs) while modern telephone switches use a combination of space- and time-division switching. In other words, each voice channel is represented by a time slot (say 1 or 2) on a physical wire pair (A or B). In order to connect two voice channels (say A1 and B2) together, the telephone switch interchanges the information between A1 and B2. It switches both the time slot and physical connection. To do this, it exchanges data between the time slots and connections 8,000 times per second, under control of digital logic that cycles through electronic lists of the current connections. Using both types of switching makes a modern switch far smaller than either a space or time switch could be by itself.

The structure of a switch is an odd number of layers of smaller, simpler subswitches. Each layer is interconnected by a web of wires that goes from each subswitch, to a set of the next layer of subswitches. In some designs, a physical (space) switching layer alternates with a time switching layer. The layers are symmetric, because in a telephone system callers can also be called. Other designs use time-switching only, throughout the switch.

A time-division subswitch reads a complete cycle of time slots into a memory, and then writes it out in a different order, also under control of a cyclic computer memory. This causes some delay in the signal.

A space-division subswitch switches electrical paths, often using some variant of a nonblocking minimal spanning switch, or a crossover switch.

Switch Control Algorithms

Fully Connected Mesh Network

One way is to have enough switching fabric to assure that the pairwise allocation will always succeed by building a fully connected mesh network. This is the method usually used in central office switches, which have low utilization of their resources.

Clos's Nonblocking Switch Algorithm

The connections between layers of subswitches of telephone switching system are scarce resources, allocated by special control logic in a fault tolerant manner. Clos networks are often used.

Fault Tolerance

Composite switches are inherently fault-tolerant. If a subswitch fails, the controlling computer can sense it during a periodic test. The computer marks all the connections to the subswitch as "in use". This prevents new calls, and does not interrupt old calls that remain working. As calls in progress end, the subswitch becomes unused, and new calls avoid the subswitch because it's already "in use." Some time later, a technician can replace the circuit board. When the next test succeeds, the connections to the repaired subsystem are marked "not in use", and the switch returns to full operation.

To prevent frustration with unsensed failures, all the connections between layers in the switch are allocated using first-in-first-out lists (queues). As a result, if a connection is faulty or noisy and the customer hangs up and redials, they will get a different set of connections and subswitches. A last-in-first-out (stack) allocation of connections might cause a continuing string of very frustrating failures.

Fire and Disaster Recovery

Telephone Exchange fire.

The central exchange, due to the system's design, is almost always a single point of failure for local calls. As the capacity of individual switches and the optical fibre which interconnects them increases, potential disruption caused by destruction of one local office will only be magnified. Multiple fibre connections can be used to provide redundancy to voice and data connections between switching centres, but careful network design is required to avoid situations where a main fibre and its backup both go through the same damaged central office as a potential common mode failure.

TERMINAL TELECOMMUNICATION

A terminal is an electronic communication hardware device that handles the input and display of data.

A terminal may be a PC or workstation connected to a network, Voice over Internet Protocol (VOIP) network endpoint, mobile data terminal such as a telematics device, or a text terminal, or textual language interface.

Terminals vary by required data type and format. Early terminals resembled typewriters. Current versions include input keyboards and output display screens.

Terminals are divided into the following three classes, according to their processing power:

- Intelligent Terminal: Includes main memory and CPU.
- Smart Terminal (fat client): Equipped with robust data processing power, but has fewer processing capabilities than an intelligent terminal.
- Dumb Terminal (thin client): Relies on the host for processing.

References

- Network-topology, definition: searchnetworking.techtarget.com, Retrieved 12 April, 2019
- ATIS committee PRQC. "network topology". ATIS Telecom Glossary 2007. Alliance for Telecommunications Industry Solutions. Retrieved 2008-10-10
- Network-topology-types, computer-networks: studytonight.com, Retrieved 13 May, 2019
- Davies, Howard; Bressan, Beatrice, eds. (2010). A history of international research networking: the people who made it happen. John Wiley & Sons. p. 2. ISBN 978-3527327102
- Switch, definition: searchnetworking.techtarget.com, Retrieved 14 June, 2019
- "Welcome hunreal.com - bluehost.com". Hunreal.com. Archived from the original on 2012-03-16. Retrieved 2012-07-01

4

Multiplexing and Multiple Access Techniques

Multiplexing is a method by which multiple digital signals are combined into one signal over a shared medium. The techniques which allow multiple mobile users to share the allotted spectrum in the most efficient manner are known as multiple access techniques. The topics elaborated in this chapter will help in gaining a better perspective about multiplexing and multiple access techniques.

FREQUENCY-DIVISION MULTIPLEXING

Frequency-Division Multiplexing (FDM) is a scheme in which numerous signals are combined for transmission on a single communications line or channel. It is analog multiplexing technique. Each signal is assigned a different frequency (sub channel) within the main channel. its requires channel synchronization. FDM multiplexing technique is based on orthogonality of sinusoids.

FDM requires that the bandwidth of a link should be greater than the combined bandwidths of the various signals to be transmitted. Thus each signal having different frequency forms a particular logical channel on the link and follows this channel only. These channels are then separated by the strips of unused bandwidth called guard bands. These guard bands prevent the signals from overlapping as shown in figure.

In FDM, signals to be transmitted must be analog signals. Thus digital signals need to be converted to analog form, if they are to use FDM.

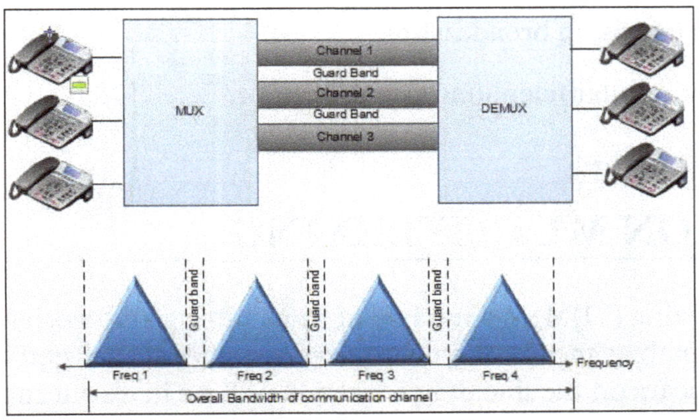

Frequency-division multiplexing.

A typical analog Internet connection via a twisted pair telephone line requires approximately three kilohertz (3 kHz) of bandwidth for accurate and reliable data transfer.

Twisted-pair lines are common in households and small businesses. But major telephone cables, operating between large businesses, government agencies, and municipalities, are capable of much larger bandwidths.

Advantages of FDM

- A large number of signals (channels) can be transmitted simultaneously.
- FDM does not need synchronization between its transmitter and receiver for proper operation.
- Demodulation of FDM is easy.
- Due to slow narrow band fading only a single channel gets affected.

Disadvantages of FDM

- The communication channel must have a very large bandwidth.
- Intermodulation distortion takes place.
- Large number of modulators and filters are required.
- FDM suffers from the problem of crosstalk.
- All the FDM channels get affected due to wideband fading.

Applications of FDM

- FDM is used for FM & AM radio broadcasting. Each AM and FM radio station uses a different carrier frequency. In AM broadcasting, these frequencies use a special band from 530 to 1700 KHz. All these signals/frequencies are multiplexed and are transmitted in air. A receiver receives all these signals but tunes only one which is required. Similarly FM broadcasting uses a bandwidth of 88 to 108 MHz.
- FDM is used in television broadcasting.
- First generation cellular telephone also uses FDM.

TIME-DIVISION MULTIPLEXING

Time-division multiplexing (TDM) is a method of transmitting and receiving independent signals over a common signal path by means of synchronized switches at each end of the transmission line so that each signal appears on the line only a fraction of time in an alternating pattern. It is used when the bit rate of the transmission medium exceeds that of the signal to be transmitted. This form of signal multiplexing was developed in telecommunications for telegraphy systems in the late 19th century, but found its most common application in digital telephony in the second half of the 20th century.

Technology

Time-division multiplexing is used primarily for digital signals, but may be applied in analog multiplexing in which two or more signals or bit streams are transferred appearing simultaneously as sub-channels in one communication channel, but are physically taking turns on the channel. The time domain is divided into several recurrent *time slots* of fixed length, one for each sub-channel. A sample byte or data block of sub-channel 1 is transmitted during time slot 1, sub-channel 2 during time slot 2, etc. One TDM frame consists of one time slot per sub-channel plus a synchronization channel and sometimes error correction channel before the synchronization. After the last sub-channel, error correction, and synchronization, the cycle starts all over again with a new frame, starting with the second sample, byte or data block from sub-channel 1, etc.

Application Examples

- The plesiochronous digital hierarchy (PDH) system, also known as the PCM system, for digital transmission of several telephone calls over the same four-wire copper cable (T-carrier or E-carrier) or fiber cable in the circuit switched digital telephone network.
- The synchronous digital hierarchy (SDH)/synchronous optical networking (SONET) network transmission standards that have replaced PDH.
- The Basic Rate Interface and Primary Rate Interface for the Integrated Services Digital Network (ISDN).
- The RIFF (WAV) audio standard interleaves left and right stereo signals on a per-sample basis.

TDM can be further extended into the time-division multiple access (TDMA) scheme, where several stations connected to the same physical medium, for example sharing the same frequency channel, can communicate. Application examples include:

- The GSM telephone system.
- The Tactical Data Links Link 16 and Link 22.

Multiplexed Digital Transmission

In circuit-switched networks, such as the public switched telephone network (PSTN), it is desirable to transmit multiple subscriber calls over the same transmission medium to effectively utilize the bandwidth of the medium. TDM allows transmitting and receiving telephone switches to create channels (*tributaries*) within a transmission stream. A standard DS0 voice signal has a data bit rate of 64 kbit/s. A TDM circuit runs at a much higher signal bandwidth, permitting the bandwidth to be divided into time frames (time slots) for each voice signal which is multiplexed onto the line by the transmitter. If the TDM frame consists of n voice frames, the line bandwidth is n 64 kbit/s.

Each voice time slot in the TDM frame is called a channel. In European systems, standard TDM frames contain 30 digital voice channels (E1), and in American systems (T1), they contain 24 channels. Both standards also contain extra bits (or bit time slots) for signaling and synchronization bits.

Multiplexing more than 24 or 30 digital voice channels is called *higher order multiplexing*. Higher order multiplexing is accomplished by multiplexing the standard TDM frames. For example, a European 120 channel TDM frame is formed by multiplexing four standard 30 channel TDM frames. At each higher order multiplex, four TDM frames from the immediate lower order are combined, creating multiplexes with a bandwidth of $n64$ kbit/s, where $n = 120, 480, 1920$, etc.

Telecommunications Systems

There are three types of synchronous TDM: T1, SONET/SDH, and ISDN.

Plesiochronous digital hierarchy (PDH) was developed as a standard for multiplexing higher order frames. PDH created larger numbers of channels by multiplexing the standard Europeans 30 channel TDM frames. This solution worked for a while; however PDH suffered from several inherent drawbacks which ultimately resulted in the development of the Synchronous Digital Hierarchy (SDH). The requirements which drove the development of SDH were these:

- Be synchronous – All clocks in the system must align with a reference clock.
- Be service-oriented – SDH must route traffic from End Exchange to End Exchange without worrying about exchanges in between, where the bandwidth can be reserved at a fixed level for a fixed period of time.
- Allow frames of any size to be removed or inserted into an SDH frame of any size.
- Easily manageable with the capability of transferring management data across links.
- Provide high levels of recovery from faults.
- Provide high data rates by multiplexing any size frame, limited only by technology.
- Give reduced bit rate errors.

SDH has become the primary transmission protocol in most PSTN networks. It was developed to allow streams 1.544 Mbit/s and above to be multiplexed, in order to create larger SDH frames known as Synchronous Transport Modules (STM). The STM-1 frame consists of smaller streams that are multiplexed to create a 155.52 Mbit/s frame. SDH can also multiplex packet based frames e.g. Ethernet, PPP and ATM.

While SDH is considered to be a transmission protocol (Layer 1 in the OSI Reference Model), it also performs some switching functions, as stated in the third bullet point requirement listed above. The most common SDH Networking functions are these:

- SDH Crossconnect: The SDH Crossconnect is the SDH version of a Time-Space-Time crosspoint switch. It connects any channel on any of its inputs to any channel on any of its outputs. The SDH Crossconnect is used in Transit Exchanges, where all inputs and outputs are connected to other exchanges.

- SDH Add-Drop Multiplexer: The SDH Add-Drop Multiplexer (ADM) can add or remove any multiplexed frame down to 1.544Mb. Below this level, standard TDM can be performed. SDH ADMs can also perform the task of an SDH Crossconnect and are used in End Exchanges where the channels from subscribers are connected to the core PSTN network.

SDH network functions are connected using high-speed optic fibre. Optic fibre uses light pulses to transmit data and is therefore extremely fast. Modern optic fibre transmission makes use of wavelength-division multiplexing (WDM) where signals transmitted across the fibre are transmitted at different wavelengths, creating additional channels for transmission. This increases the speed and capacity of the link, which in turn reduces both unit and total costs.

Statistical Time-division Multiplexing

Statistical time-division multiplexing (STDM) is an advanced version of TDM in which both the address of the terminal and the data itself are transmitted together for better routing. Using STDM allows bandwidth to be split over one line. Many college and corporate campuses use this type of TDM to distribute bandwidth.

On a 10-Mbit line entering a network, STDM can be used to provide 178 terminals with a dedicated 56k connection (178 × 56k = 9.96Mb). A more common use however is to only grant the bandwidth when that much is needed. STDM does not reserve a time slot for each terminal, rather it assigns a slot when the terminal is requiring data to be sent or received.

In its primary form, TDM is used for circuit mode communication with a fixed number of channels and constant bandwidth per channel. Bandwidth reservation distinguishes time-division multiplexing from statistical multiplexing such as statistical time-division multiplexing. In pure TDM, the time slots are recurrent in a fixed order and pre-allocated to the channels, rather than scheduled on a packet-by-packet basis.

In dynamic TDMA, a scheduling algorithm dynamically reserves a variable number of time slots in each frame to variable bit-rate data streams, based on the traffic demand of each data stream. Dynamic TDMA is used in:

- HIPERLAN/2.
- Dynamic synchronous transfer mode.
- IEEE 802.16a.

Asynchronous time-division multiplexing (ATDM), is an alternative nomenclature in which STDM designates synchronous time-division multiplexing, the older method that uses fixed time slots.

FREQUENCY-DIVISION MULTIPLE ACCESS

Frequency Division Multiple Access (FDMA) is one of the most common analogue multiple access methods. The frequency band is divided into channels of equal bandwidth so that each conversation is carried on a different frequency.

In FDMA method, guard bands are used between the adjacent signal spectra to minimize cross-talk between the channels. A specific frequency band is given to one person, and it will received by identifying each of the frequency on the receiving end. It is often used in the first generation of analog mobile phone.

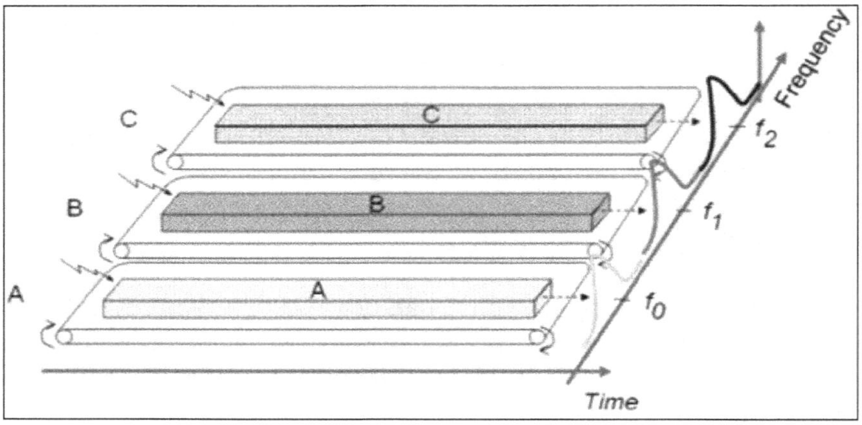

Advantages of FDMA

As FDMA systems use low bit rates (large symbol time) compared to average delay spread, it offers the following advantages –

- Reduces the bit rate information and the use of efficient numerical codes increases the capacity.
- It reduces the cost and lowers the inter symbol interference (ISI).
- Equalization is not necessary.
- An FDMA system can be easily implemented. A system can be configured so that the improvements in terms of speech encoder and bit rate reduction may be easily incorporated.
- Since the transmission is continuous, less number of bits are required for synchronization and framing.

Disadvantages of FDMA

Although FDMA offers several advantages, it has a few drawbacks as well, which are listed below:

- It does not differ significantly from analog systems; improving the capacity depends on the signal-to-interference reduction, or a signal-to-noise ratio (SNR).
- The maximum flow rate per channel is fixed and small.
- Guard bands lead to a waste of capacity.
- Hardware implies narrowband filters, which cannot be realized in VLSI and therefore increases the cost.

TIME-DIVISION MULTIPLE ACCESS

Time-division multiple access (TDMA) is a channel access method for shared-medium networks. It allows several users to share the same frequency channel by dividing the signal into different time slots. The users transmit in rapid succession, one after the other, each using its own time slot. This allows multiple stations to share the same transmission medium (e.g. radio frequency channel) while using only a part of its channel capacity. TDMA is used in the digital 2G cellular systems such as Global System for Mobile Communications (GSM), IS-136, Personal Digital Cellular (PDC) and iDEN, and in the Digital Enhanced Cordless Telecommunications (DECT) standard for portable phones. TDMA was first used in satellite communication systems by Western Union in its Westar 3 communications satellite in 1979. It is now used extensively in satellite communications, combat-net radio systems, and passive optical network (PON) networks for upstream traffic from premises to the operator.

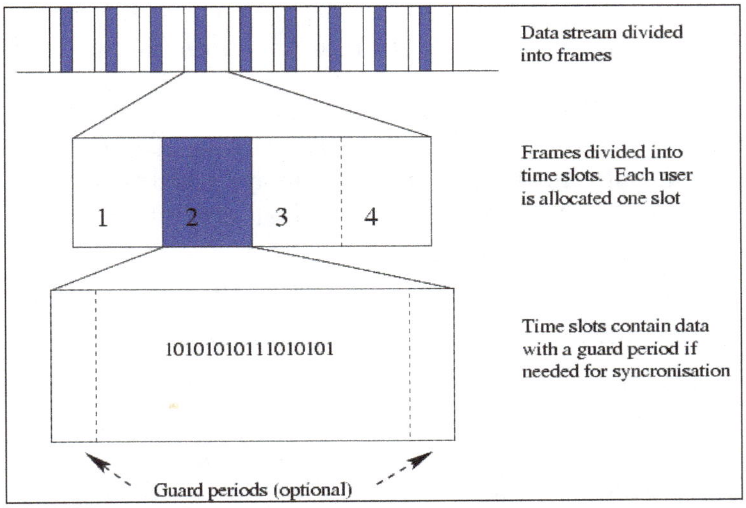

TDMA frame structure showing a data stream divided into frames and those frames divided into time slots.

TDMA is a type of time-division multiplexing (TDM), with the special point that instead of having one transmitter connected to one receiver, there are multiple transmitters. In the case of the *uplink* from a mobile phone to a base station this becomes particularly difficult because the mobile phone can move around and vary the *timing advance* required to make its transmission match the gap in transmission from its peers.

TDMA Characteristics

- Shares single carrier frequency with multiple users.
- Non-continuous transmission makes handoff simpler.
- Slots can be assigned on demand in dynamic TDMA.
- Less stringent power control than CDMA due to reduced intra cell interference.
- Higher synchronization overhead than CDMA.

- Advanced equalization may be necessary for high data rates if the channel is "frequency selective" and creates Intersymbol interference.
- Cell breathing (borrowing resources from adjacent cells) is more complicated than in CDMA.
- Frequency/slot allocation complexity.
- Pulsating power envelope: interference with other devices.

TDMA in Mobile Phone Systems

2G Systems

Most 2G cellular systems, with the notable exception of IS-95, are based on TDMA. GSM, D-AMPS, PDC, iDEN, and PHS are examples of TDMA cellular systems. GSM combines TDMA with Frequency Hopping and wideband transmission to minimize common types of interference.

In the GSM system, the synchronization of the mobile phones is achieved by sending timing advance commands from the base station which instructs the mobile phone to transmit earlier and by how much. This compensates for the propagation delay resulting from the light speed velocity of radio waves. The mobile phone is not allowed to transmit for its entire time slot, but there is a guard interval at the end of each time slot. As the transmission moves into the guard period, the mobile network adjusts the timing advance to synchronize the transmission.

Initial synchronization of a phone requires even more care. Before a mobile transmits there is no way to actually know the offset required. For this reason, an entire time slot has to be dedicated to mobiles attempting to contact the network; this is known as the random-access channel (RACH) in GSM. The mobile attempts to broadcast at the beginning of the time slot, as received from the network. If the mobile is located next to the base station, there will be no time delay and this will succeed. If, however, the mobile phone is at just less than 35 km from the base station, the time delay will mean the mobile's broadcast arrives at the very end of the time slot. In that case, the mobile will be instructed to broadcast its messages starting nearly a whole time slot earlier than would be expected otherwise. Finally, if the mobile is beyond the 35 km cell range in GSM, then the RACH will arrive in a neighbouring time slot and be ignored. It is this feature, rather than limitations of power, that limits the range of a GSM cell to 35 km when no special extension techniques are used. By changing the synchronization between the uplink and downlink at the base station, however, this limitation can be overcome.

3G Systems

Although most major 3G systems are primarily based upon CDMA, time-division duplexing (TDD), packet scheduling (dynamic TDMA) and packet oriented multiple access schemes are available in 3G form, combined with CDMA to take advantage of the benefits of both technologies.

While the most popular form of the UMTS 3G system uses CDMA and frequency division duplexing (FDD) instead of TDMA, TDMA is combined with CDMA and time-division duplexing in two standard UMTS UTRA.

TDMA in Wired Networks

The ITU-T G.hn standard, which provides high-speed local area networking over existing home wiring (power lines, phone lines and coaxial cables) is based on a TDMA scheme. In G.hn, a "master" device allocates "Contention-Free Transmission Opportunities" (CFTXOP) to other "slave" devices in the network. Only one device can use a CFTXOP at a time, thus avoiding collisions. FlexRay protocol which is also a wired network used for safety-critical communication in modern cars, uses the TDMA method for data transmission control.

Comparison with other Multiple-access Schemes

In radio systems, TDMA is usually used alongside frequency-division multiple access (FDMA) and frequency division duplex (FDD); the combination is referred to as FDMA/TDMA/FDD. This is the case in both GSM and IS-136 for example. Exceptions to this include the DECT and Personal Handy-phone System (PHS) micro-cellular systems, UMTS-TDD UMTS variant, and China's TD-SCDMA, which use time-division duplexing, where different time slots are allocated for the base station and handsets on the same frequency.

A major advantage of TDMA is that the radio part of the mobile only needs to listen and broadcast for its own time slot. For the rest of the time, the mobile can carry out measurements on the network, detecting surrounding transmitters on different frequencies. This allows safe inter frequency handovers, something which is difficult in CDMA systems, not supported at all in IS-95 and supported through complex system additions in Universal Mobile Telecommunications System (UMTS). This in turn allows for co-existence of microcell layers with macrocell layers.

CDMA, by comparison, supports "soft hand-off" which allows a mobile phone to be in communication with up to 6 base stations simultaneously, a type of "same-frequency handover". The incoming packets are compared for quality, and the best one is selected. CDMA's "cell breathing" characteristic, where a terminal on the boundary of two congested cells will be unable to receive a clear signal, can often negate this advantage during peak periods.

A disadvantage of TDMA systems is that they create interference at a frequency which is directly connected to the time slot length. This is the buzz which can sometimes be heard if a TDMA phone is left next to a radio or speakers. Another disadvantage is that the "dead time" between time slots limits the potential bandwidth of a TDMA channel. These are implemented in part because of the difficulty in ensuring that different terminals transmit at exactly the times required. Handsets that are moving will need to constantly adjust their timings to ensure their transmission is received at precisely the right time, because as they move further from the base station, their signal will take longer to arrive. This also means that the major TDMA systems have hard limits on cell sizes in terms of range, though in practice the power levels required to receive and transmit over distances greater than the supported range would be mostly impractical anyway.

Dynamic TDMA

In dynamic time-division multiple access (dynamic TDMA), a scheduling algorithm dynamically reserves a variable number of time slots in each frame to variable bit-rate data streams, based on the traffic demand of each data stream. Dynamic TDMA is used in:

- HIPERLAN/2 broadband radio access network.
- IEEE 802.16a WiMax.
- Bluetooth.
- Military Radios/Tactical Data Link.
- TD-SCDMA.
- ITU-T G.hn.
- Simulation of TDMA/DTMA links.

POLARIZATION-DIVISION MULTIPLEXING

Polarization-division multiplexing (PDM) is a physical layer method for multiplexing signals carried on electromagnetic waves, allowing two channels of information to be transmitted on the same carrier frequency by using waves of two orthogonal polarization states. It is used in microwave links such as satellite television downlinks to double the bandwidth by using two orthogonally polarized feed antennas in satellite dishes. It is also used in fiber optic communication by transmitting separate left and right circularly polarized light beams through the same optical fiber.

Radio

Polarization techniques have long been used in radio transmission to reduce interference between channels, particularly at VHF frequencies and beyond.

Under some circumstances, the data rate of a radio link can be doubled by transmitting two separate channels of radio waves on the same frequency, using orthogonal polarization. For example, in point to point terrestrial microwave links, the transmitting antenna can have two feed antennas; a vertical feed antenna which transmits microwaves with their electric field vertical (vertical polarization), and a horizontal feed antenna which transmits microwaves on the same frequency with their electric field horizontal (horizontal polarization). These two separate channels can be received by vertical and horizontal feed antennas at the receiving station. For satellite communications, orthogonal circular polarization is often used instead, (i.e. right- and left-handed), as the sense of circular polarization is not changed by the relative orientation of the antenna in space.

A dual polarization system comprises usually two independent transmitters, each of which can be connected by means of waveguide or TEM lines (such as coaxial cables or stripline or quasi-TEM such as microstrip) to a single-polarization antenna for its standard operation. Although two separate single-polarization antennas can be used for PDM (or two adjacent feeds in a reflector antenna), radiating two independent polarization states can be often easily achieved by means of a single dual-polarization antenna.

When the transmitter has a waveguide interface, typically rectangular in order to be in single-mode region at the operating frequency, a dual-polarized antenna with a circular (or square) waveguide port is the radiating element chosen for modern communication systems. The circular or square waveguide port is needed so that at least two degenerate modes are supported. An ad-hoc component must be therefore introduced in such situations to merge two separate single-polarized signals into one dual-polarized physical interface, namely an ortho-mode transducer (OMT).

In case the transmitter has TEM or quasi-TEM output connections, instead, a dual-polarization antenna often presents separate connections (i.e. a printed square patch antenna with two feed points), and embeds the function of an OMT by means of intrinsically transferring the two excitation signals to the orthogonal polarization states.

A dual-polarized signal thus carries two independent data streams to a receiving antenna, which can itself be a single-polarized one, for receiving only one of the two streams at a time, or a dual-polarized model, again relaying its received signal to two single-polarization output connectors (via an OMT if in waveguide).

The ideal dual-polarization system lies its foundation onto the perfect orthogonality of the two polarization states, and any of the single-polarized interfaces at the receiver would theoretically contain only the signal meant to be transmitted by the desired polarization, thus introducing no interference and allowing the two data streams to be multiplexed and demultiplexed transparently without any degradation due to the coexistence with the other.

Companies working on commercial PDM technology include Siae Microelettronica, Huawei and Alcatel-Lucent. Some types of outdoor microwave radios have integrated orthomode transducers and operate in both polarities from a single radio unit, performing cross-polarization interference cancellation (XPIC) within the radio unit itself. Alternatively, the orthomode transducer may be built into the antenna, and allow connection of separate radios, or separate ports of the same radio, to the antenna.

CableFree 2+0 XPIC Microwave Link showing OMT and two ODUs connected to H & V polarity ports.

Cross-Polarization Interference Cancellation (XPIC)

Practical systems, however, suffer from non-ideal behaviors which mix the signals and the polarization states together:

- The OMT at the transmitting side has a finite cross-polarization discrimination (XPD) and thus leaks part of the signals meant to be transmitted in one polarization to the other.
- The transmitting antenna has a finite XPD and thus leaks part of its input polarizations to the other radiated polarization state.
- Propagation in presence of rain, snow, hail creates depolarization, as part of the two impinging polarizations is leaked to the other.
- The finite XPD of the receiving antenna acts similarly to the transmitting side and the relative alignment of the two antennas contributes to a loss of system XPD.
- The finite XPD of the receiving OMT likewise further mixes the signals from the dual-polarized port to the single-polarized ports.

As a consequence, the signal at one of the received single-polarization terminals actually contains a dominant quantity of the desired signal (meant to be transmitted onto one polarization) and a minor amount of undesired signal (meant to be transported by the other polarization), which represents an interference over the former. As a consequence, each received signal must be cleared of the interference level in order to reach the required signal-to-noise-and-interference ratio (SNIR) needed by the receiving stages, which may be of the order of more than 30 dB for high-level M-QAM schemes. Such operation is carried out by a cross-polarization-interference cancellation (XPIC), typically implemented as a baseband digital stage.

Compared to spatial multiplexing, received signals for a PMD system have a much more favourable carrier-to-interference ratio, as the amount of leakage is often much smaller than the useful signal, whereas spatial multiplexing operates with an amount of interference equal to the amount of useful signal. This observation, valid for a good PMD design, allows the adaptive XPIC to be designed in a simpler manner than a general MIMO cancelling scheme, since the starting point (without cancellation) is typically already sufficient for establishing a low-capacity link by means of a reduced modulation.

An XPIC typically acts on one of the received signals "C" containing the desired signal as dominant term and uses the other received "X" signal too (containing the interfering signal as dominant term). The XPIC algorithm multiplies the "X" by a complex coefficient and then adds it to the received "C". The complex recombination coefficient is adjusted adaptively to maximize the MMSE as measured on the recombination. Once the MMSE is improved to the required level, the two terminals can switch to high-order modulations.

Differential Cross-Polarized Wireless Communications

Is a novel method for polarized antenna transmission utilizing a differential technique.

Photonics

Polarization-division multiplexing is typically used together with phase modulation or optical

QAM, allowing transmission speeds of 100 Gbit/s or more over a single wavelength. Sets of PDM wavelength signals can then be carried over wavelength-division multiplexing infrastructure, potentially substantially expanding its capacity. Multiple polarization signals can be combined to form new states of polarization, which is known as parallel polarization state generation.

The major problem with the practical use of PDM over fiber-optic transmission systems are the drifts in polarization state that occur continuously over time due to physical changes in the fibre environment. Over a long-distance system, these drifts accumulate progressively without limit, resulting in rapid and erratic rotation of the polarized light's Jones vector over the entire Poincaré sphere. Polarization mode dispersion, polarization-dependent loss. and cross-polarization modulation are other phenomena that can cause problems in PDM systems.

For this reason, PDM is generally used in conjunction with advanced channel coding techniques, allowing the use of digital signal processing to decode the signal in a way that is resilient to polarization-related signal artifacts. Modulations used include PDM-QPSK and PDM-DQPSK. Companies working on commercial PDM technology include Alcatel-Lucent, Ciena, Cisco Systems, Huawei and Infinera.

ORBITAL ANGULAR MOMENTUM MULTIPLEXING

Orbital angular momentum (OAM) multiplexing is a physical layer method for multiplexing signals carried on electromagnetic waves using the orbital angular momentum of the electromagnetic waves to distinguish between the different orthogonal signals.

Orbital angular momentum is one of two forms of angular momentum of light. OAM is distinct from, and should not be confused with, light spin angular momentum. The spin angular momentum of light offers only two orthogonal quantum states corresponding to the two states of circular polarization, and can be demonstrated to be equivalent to a combination of polarization multiplexing and phase shifting. OAM on the other hand relies on an extended beam of light, and the higher quantum degrees of freedom which come with the extension. OAM multiplexing can thus access a potentially unbounded set of states, and as such offer a much larger number of channels, subject only to the constraints of real-world optics.

As of 2013, although OAM multiplexing promises very significant improvements in bandwidth when used in concert with other existing modulation and multiplexing schemes, it is still an experimental technique, and has so far only been demonstrated in the laboratory. Following the early claim that OAM exploits a new quantum mode of information propagation, the technique has become controversial; however nowadays it can be understood to be a particular form of tightly modulated MIMO multiplexing strategy, obeying classical information theoretic bounds. OAM multiplexing was demonstrated using light beams in free space as early as 2004. Since then, research into OAM has proceeded in two areas: radio frequency and optical transmission.

Radio Frequency

An experiment in 2011 demonstrated OAM multiplexing of two incoherent radio signals over a distance of 442 m. It has been claimed that OAM does not improve on what can achieved with conventional linear-momentum based RF systems which already use MIMO, since theoretical work suggests that, at radio frequencies, conventional MIMO techniques can be shown to duplicate many of the linear-momentum properties of OAM-carrying radio beam, leaving little or no extra performance gain.

In November 2012, there were reports of disagreement about the basic theoretical concept of OAM multiplexing at radio frequencies between the research groups of Tamburini and Thide, and many different camps of communications engineers and physicists, with some declaring their belief that OAM multiplexing was just an implementation of MIMO, and others holding to their assertion that OAM multiplexing is a distinct, experimentally confirmed phenomenon.

In 2014, a group of researchers described an implementation of a communication link over 8 millimetre-wave channels multiplexed using a combination of OAM and polarization-mode multiplexing to achieve an aggregate bandwidth of 32 Gbit/s over a distance of 2.5 metres.

The industrial interest for long-distance microwave OAM multiplexing seems to have been diminishing since 2015, when some of the original promoters of OAM-based communication at radio frequencies have published a theoretical investigation showing that there is no real gain beyond traditional spatial multiplexing in terms of capacity and overall antenna occupation.

Optical

OAM multiplexing is used in the optical domain. In 2012, researchers demonstrated OAM-multiplexed optical transmission speeds of up to 2.5 Tbits/s using 8 distinct OAM channels in a single beam of light, but only over a very short free-space path of roughly one metre. Work is ongoing on applying OAM techniques to long-range practical free-space optical communication links.

OAM multiplexing can not be implemented in the existing long-haul optical fiber systems, since these systems are based on single-mode fibers, which inherently do not support OAM states of light. Instead, few-mode or multi-mode fibers need to be used. Additional problem for OAM multiplexing implementation is caused by the mode coupling that is present in conventional fibers, which cause changes in the spin angular momentum of modes under normal conditions and changes in orbital angular momentum when fibers are bent or stressed. Because of this mode instability, direct-detection OAM multiplexing has not yet been realized in long-haul communications. In 2012, transmission of OAM states with 97% purity after 20 meters over special fibers was demonstrated by researchers at Boston University. Later experiments have shown stable propagation of these modes over distances of 50 meters, and further improvements of this distance are the subject of ongoing work. Other ongoing research on making OAM multiplexing work over future fibre-optic transmission systems includes the possibility of using similar techniques to those used to compensate mode rotation in optical polarization multiplexing.

Alternative to direct-detection OAM multiplexing is a computationally complex coherent-detection with (MIMO) digital signal processing (DSP) approach, that can be used to achieve long-haul communication, where strong mode coupling is suggested to be beneficial for coherent-detection-based systems.

CODE-DIVISION MULTIPLE ACCESS

Code-division multiple access (CDMA) is a channel access method used by various radio communication technologies.

CDMA is an example of multiple access, where several transmitters can send information simultaneously over a single communication channel. This allows several users to share a band of frequencies. To permit this without undue interference between the users, CDMA employs spread spectrum technology and a special coding scheme (where each transmitter is assigned a code).

CDMA is used as the access method in many mobile phone standards. IS-95, also called "cdmaOne", and its 3G evolution CDMA2000, are often simply referred to as "CDMA", but UMTS, the 3G standard used by GSM carriers, also uses "wideband CDMA", or W-CDMA, as well as TD-CDMA and TD-SCDMA, as its radio technologies.

Uses

A CDMA2000 mobile phone.

- One of the early applications for code-division multiplexing is in the Global Positioning System (GPS). This predates and is distinct from its use in mobile phones.
- The Qualcomm standard IS-95, marketed as cdmaOne.
- The Qualcomm standard IS-2000, known as CDMA2000, is used by several mobile phone companies, including the Globalstar network.
- The UMTS 3G mobile phone standard, which uses W-CDMA.
- CDMA has been used in the OmniTRACS satellite system for transportation logistics.

Steps in CDMA Modulation

CDMA is a spread-spectrum multiple-access technique. A spread-spectrum technique spreads the bandwidth of the data uniformly for the same transmitted power. A spreading code is a pseudo-random code that has a narrow ambiguity function, unlike other narrow pulse codes. In CDMA a locally generated code runs at a much higher rate than the data to be transmitted. Data for transmission is combined by bitwise XOR (exclusive OR) with the faster code. The figure shows how a spread-spectrum signal is generated. The data signal with pulse duration of T_b (symbol period) is XORed with the code signal with pulse duration of T_c (chip period). (bandwidth is proportional to $1/T$, where T = bit time.) Therefore, the bandwidth of the data signal is $1/T_b$ and the bandwidth of the spread spectrum signal is $1/T_c$. Since T_c is much smaller than T_b, the bandwidth of the spread-spectrum signal is much larger than the bandwidth of the original signal. The ratio T_b/T_c is called the spreading factor or processing gain and determines to a certain extent the upper limit of the total number of users supported simultaneously by a base station.

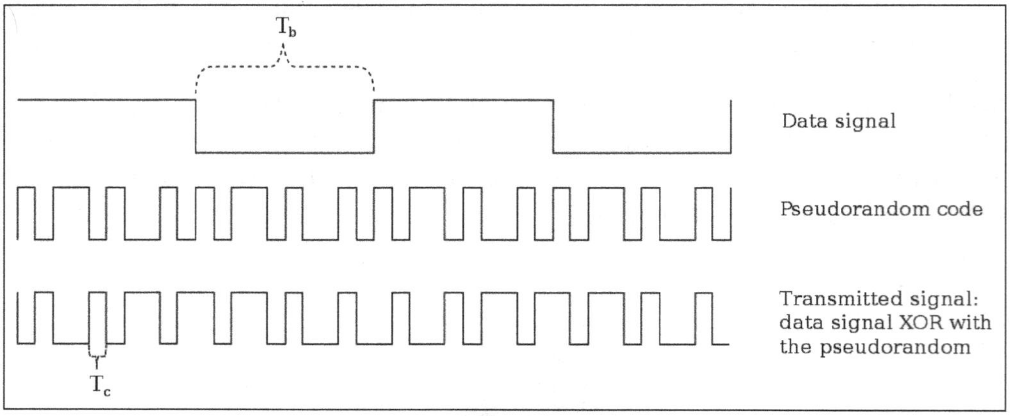

Generation of a CDMA signal.

Each user in a CDMA system uses a different code to modulate their signal. Choosing the codes used to modulate the signal is very important in the performance of CDMA systems. The best performance occurs when there is good separation between the signal of a desired user and the signals of other users. The separation of the signals is made by correlating the received signal with the locally generated code of the desired user. If the signal matches the desired user's code, then the correlation function will be high and the system can extract that signal. If the desired user's code has nothing in common with the signal, the correlation should be as close to zero as possible (thus eliminating the signal); this is referred to as cross-correlation. If the code is correlated with the signal at any time offset other than zero, the correlation should be as close to zero as possible. This is referred to as auto-correlation and is used to reject multi-path interference.

An analogy to the problem of multiple access is a room (channel) in which people wish to talk to each other simultaneously. To avoid confusion, people could take turns speaking (time division), speak at different pitches (frequency division), or speak in different languages (code division). CDMA is analogous to the last example where people speaking the same language can understand each other, but other languages are perceived as noise and rejected. Similarly, in radio CDMA, each group of users is given a shared code. Many codes occupy the same channel, but only users associated with a particular code can communicate.

In general, CDMA belongs to two basic categories: synchronous (orthogonal codes) and asynchronous (pseudorandom codes).

Code-division Multiplexing (Synchronous CDMA)

The digital modulation method is analogous to those used in simple radio transceivers. In the analog case, a low-frequency data signal is time-multiplied with a high-frequency pure sine-wave carrier and transmitted. This is effectively a frequency convolution (Wiener–Khinchin theorem) of the two signals, resulting in a carrier with narrow sidebands. In the digital case, the sinusoidal carrier is replaced by Walsh functions. These are binary square waves that form a complete orthonormal set. The data signal is also binary and the time multiplication is achieved with a simple XOR function. This is usually a Gilbert cell mixer in the circuitry.

Synchronous CDMA exploits mathematical properties of orthogonality between vectors representing the data strings. For example, binary string *1011* is represented by the vector (1, 0, 1, 1). Vectors can be multiplied by taking their dot product, by summing the products of their respective components (for example, if u = (a, b) and v = (c, d), then their dot product u·v = $ac + bd$). If the dot product is zero, the two vectors are said to be *orthogonal* to each other. Some properties of the dot product aid understanding of how W-CDMA works. If vectors a and b are orthogonal, then $\mathbf{a} \cdot \mathbf{b} = 0$ and:

$$\mathbf{a} \cdot (\mathbf{a} + \mathbf{b}) = \|\mathbf{a}\|^2, \text{ since } \mathbf{a} \cdot \mathbf{a} + \mathbf{a} \cdot \mathbf{b} = \|\mathbf{a}\|^2 + 0,$$

$$\mathbf{a} \cdot (-\mathbf{a} + \mathbf{b}) = -\|\mathbf{a}\|^2, \text{ since } -\mathbf{a} \cdot \mathbf{a} + \mathbf{a} \cdot \mathbf{b} = -\|\mathbf{a}\|^2 + 0,$$

$$\mathbf{b} \cdot (\mathbf{a} + \mathbf{b}) = \|\mathbf{b}\|^2, \text{ since } \mathbf{b} \cdot \mathbf{a} + \mathbf{b} \cdot \mathbf{b} = 0 + \|\mathbf{b}\|^2,$$

$$\mathbf{b} \cdot (\mathbf{a} - \mathbf{b}) = -\|\mathbf{b}\|^2, \text{ since } \mathbf{b} \cdot \mathbf{a} - \mathbf{b} \cdot \mathbf{b} = 0 - \|\mathbf{b}\|^2.$$

Each user in synchronous CDMA uses a code orthogonal to the others' codes to modulate their signal. An example of 4 mutually orthogonal digital signals is shown in the figure below. Orthogonal codes have a cross-correlation equal to zero; in other words, they do not interfere with each other. In the case of IS-95, 64-bit Walsh codes are used to encode the signal to separate different users. Since each of the 64 Walsh codes is orthogonal to all other, the signals are channelized into 64 orthogonal signals. The following example demonstrates how each user's signal can be encoded and decoded.

Example:

Start with a set of vectors that are mutually orthogonal. (Although mutual orthogonality is the only condition, these vectors are usually constructed for ease of decoding, for example columns or rows from Walsh matrices.) An example of orthogonal functions is shown in the adjacent picture. These vectors will be assigned to individual users and are called the *code*, *chip code*, or *chipping code*. In the interest of brevity, the rest of this example uses codes v with only two bits.

Each user is associated with a different code, say v. A 1 bit is represented by transmitting a positive code v, and a 0 bit is represented by a negative code −v. For example, if v = (v_0, v_1) = (1, −1) and the data that the user wishes to transmit is (1, 0, 1, 1), then the transmitted symbols would be

$$(\mathrm{v}, -\mathrm{v}, \mathrm{v}, \mathrm{v}) = (v_0, v_1, -v_0, -v_1, v_0, v_1, v_0, v_1) = (1, -1, -1, 1, 1, -1, 1, -1).$$

we call this constructed vector the *transmitted vector*.

Each sender has a different, unique vector v chosen from that set, but the construction method of the transmitted vector is identical.

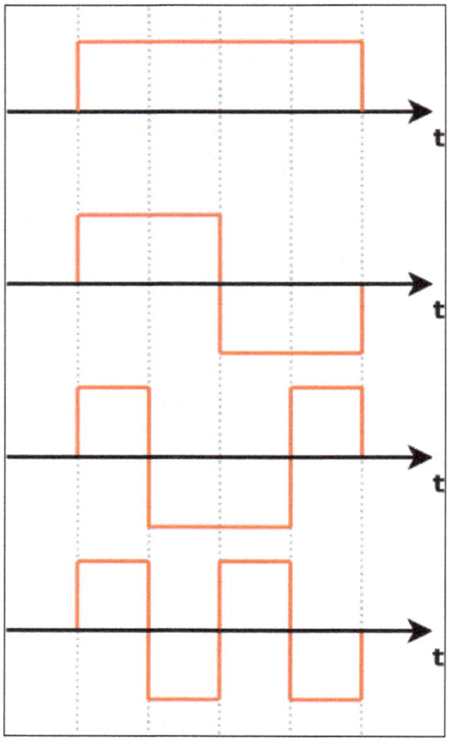

An example of 4 mutually orthogonal digital signals.

Now, due to physical properties of interference, if two signals at a point are in phase, they add to give twice the amplitude of each signal, but if they are out of phase, they subtract and give a signal that is the difference of the amplitudes. Digitally, this behaviour can be modelled by the addition of the transmission vectors, component by component.

If sender0 has code (1, −1) and data (1, 0, 1, 1), and sender1 has code (1, 1) and data (0, 0, 1, 1), and both senders transmit simultaneously, then this table describes the coding steps:

Step	Encode sender0	Encode sender1
0	code0 = (1, −1), data0 = (1, 0, 1, 1)	code1 = (1, 1), data1 = (0, 0, 1, 1)
1	encode0 = 2(1, 0, 1, 1) − (1, 1, 1, 1) = (1, −1, 1, 1)	encode1 = 2(0, 0, 1, 1) − (1, 1, 1, 1) = (−1, −1, 1, 1)
2	signal0 = encode0 ⊗ code0 = (1, −1, 1, 1) ⊗ (1, −1) = (1, −1, −1, 1, 1, −1, 1, −1)	signal1 = encode1 ⊗ code1 = (−1, −1, 1, 1) ⊗ (1, 1) = (−1, −1, −1, −1, 1, 1, 1, 1)

Because signal0 and signal1 are transmitted at the same time into the air, they add to produce the raw signal:

$$(1, -1, -1, 1, 1, -1, 1, -1) + (-1, -1, -1, -1, 1, 1, 1, 1) = (0, -2, -2, 0, 2, 0, 2, 0).$$

This raw signal is called an interference pattern. The receiver then extracts an intelligible signal for any known sender by combining the sender's code with the interference pattern. The following table explains how this works and shows that the signals do not interfere with one another:

Step	Decode sender0	Decode sender1
0	code0 = (1, −1), signal = (0, −2, −2, 0, 2, 0, 2, 0)	code1 = (1, 1), signal = (0, −2, −2, 0, 2, 0, 2, 0)
1	decode0 = pattern.vector0	decode1 = pattern.vector1
2	decode0 = ((0, −2), (−2, 0), (2, 0), (2, 0))·(1, −1)	decode1 = ((0, −2), (−2, 0), (2, 0), (2, 0))·(1, 1)
3	decode0 = ((0 + 2), (−2 + 0), (2 + 0), (2 + 0))	decode1 = ((0 − 2), (−2 + 0), (2 + 0), (2 + 0))
4	data0=(2, −2, 2, 2), meaning (1, 0, 1, 1)	data1=(−2, −2, 2, 2), meaning (0, 0, 1, 1)

Further, after decoding, all values greater than 0 are interpreted as 1, while all values less than zero are interpreted as 0. For example, after decoding, data0 is (2, −2, 2, 2), but the receiver interprets this as (1, 0, 1, 1). Values of exactly 0 means that the sender did not transmit any data, as in the following example:

Assume signal0 = (1, −1, −1, 1, 1, −1, 1, −1) is transmitted alone. The following table shows the decode at the receiver:

Step	Decode sender0	Decode sender1
0	code0 = (1, −1), signal = (1, −1, −1, 1, 1, −1, 1, −1)	code1 = (1, 1), signal = (1, −1, −1, 1, 1, −1, 1, −1)
1	decode0 = pattern.vector0	decode1 = pattern.vector1
2	decode0 = ((1, −1), (−1, 1), (1, −1), (1, −1))·(1, −1)	decode1 = ((1, −1), (−1, 1), (1, −1), (1, −1))·(1, 1)
3	decode0 = ((1 + 1), (−1 − 1), (1 + 1), (1 + 1))	decode1 = ((1 − 1), (−1 + 1), (1 − 1), (1 − 1))
4	data0 = (2, −2, 2, 2), meaning (1, 0, 1, 1)	data1 = (0, 0, 0, 0), meaning no data

When the receiver attempts to decode the signal using sender1's code, the data is all zeros, therefore the cross-correlation is equal to zero and it is clear that sender1 did not transmit any data.

Asynchronous CDMA

When mobile-to-base links cannot be precisely coordinated, particularly due to the mobility of the handsets, a different approach is required. Since it is not mathematically possible to create signature sequences that are both orthogonal for arbitrarily random starting points and which make full use of the code space, unique "pseudo-random" or "pseudo-noise" (PN) sequences are used in asynchronous CDMA systems. A PN code is a binary sequence that appears random but can be reproduced in a deterministic manner by intended receivers. These PN codes are used to encode and decode a user's signal in asynchronous CDMA in the same manner as the orthogonal codes in synchronous CDMA. These PN sequences are statistically uncorrelated, and the sum of a large number of PN sequences results in multiple access interference (MAI) that is approximated by a Gaussian noise process (following the central limit theorem in statistics). Gold codes are an example of a PN suitable for this purpose, as there is low correlation between the codes. If all of the users are received with the same power level, then the variance (e.g., the noise power) of the MAI increases in direct proportion to the number of users. In other words, unlike synchronous CDMA, the signals of other users will appear as noise to the signal of interest and interfere slightly with the desired signal in proportion to number of users.

All forms of CDMA use spread-spectrum process gain to allow receivers to partially discriminate against unwanted signals. Signals encoded with the specified PN sequence (code) are received, while signals with different codes (or the same code but a different timing offset) appear as wideband noise reduced by the process gain.

Since each user generates MAI, controlling the signal strength is an important issue with CDMA transmitters. A CDM (synchronous CDMA), TDMA, or FDMA receiver can in theory completely reject arbitrarily strong signals using different codes, time slots or frequency channels due to the orthogonality of these systems. This is not true for asynchronous CDMA; rejection of unwanted signals is only partial. If any or all of the unwanted signals are much stronger than the desired signal, they will overwhelm it. This leads to a general requirement in any asynchronous CDMA system to approximately match the various signal power levels as seen at the receiver. In CDMA cellular, the base station uses a fast closed-loop power-control scheme to tightly control each mobile's transmit power.

Advantages of Asynchronous CDMA Over other Techniques

Efficient Practical Utilization of the Fixed Frequency Spectrum

In theory CDMA, TDMA and FDMA have exactly the same spectral efficiency, but, in practice, each has its own challenges – power control in the case of CDMA, timing in the case of TDMA, and frequency generation/filtering in the case of FDMA.

TDMA systems must carefully synchronize the transmission times of all the users to ensure that they are received in the correct time slot and do not cause interference. Since this cannot be perfectly controlled in a mobile environment, each time slot must have a guard time, which reduces the probability that users will interfere, but decreases the spectral efficiency.

Similarly, FDMA systems must use a guard band between adjacent channels, due to the unpredictable Doppler shift of the signal spectrum because of user mobility. The guard bands will reduce the probability that adjacent channels will interfere, but decrease the utilization of the spectrum.

Flexible Allocation of Resources

Asynchronous CDMA offers a key advantage in the flexible allocation of resources i.e. allocation of PN codes to active users. In the case of CDM (synchronous CDMA), TDMA, and FDMA the number of simultaneous orthogonal codes, time slots, and frequency slots respectively are fixed, hence the capacity in terms of the number of simultaneous users is limited. There are a fixed number of orthogonal codes, time slots or frequency bands that can be allocated for CDM, TDMA, and FDMA systems, which remain underutilized due to the bursty nature of telephony and packetized data transmissions. There is no strict limit to the number of users that can be supported in an asynchronous CDMA system, only a practical limit governed by the desired bit error probability since the SIR (signal-to-interference ratio) varies inversely with the number of users. In a bursty traffic environment like mobile telephony, the advantage afforded by asynchronous CDMA is that the performance (bit error rate) is allowed to fluctuate randomly, with an average value determined by the number of users times the percentage of utilization. Suppose there are $2N$ users that only talk half of the time, then $2N$ users can be accommodated with the same *average* bit error probability as N users that talk all of the time. The key difference here is that the bit error probability for N users talking all of the time is constant, whereas it is a *random* quantity (with the same mean) for $2N$ users talking half of the time.

In other words, asynchronous CDMA is ideally suited to a mobile network where large numbers of transmitters each generate a relatively small amount of traffic at irregular intervals. CDM (synchronous CDMA), TDMA, and FDMA systems cannot recover the underutilized resources inherent to bursty traffic due to the fixed number of orthogonal codes, time slots or frequency channels that can be assigned to individual transmitters. For instance, if there are *N* time slots in a TDMA system and 2*N* users that talk half of the time, then half of the time there will be more than *N* users needing to use more than *N* time slots. Furthermore, it would require significant overhead to continually allocate and deallocate the orthogonal-code, time-slot or frequency-channel resources. By comparison, asynchronous CDMA transmitters simply send when they have something to say and go off the air when they don't, keeping the same PN signature sequence as long as they are connected to the system.

Spread-spectrum Characteristics of CDMA

Most modulation schemes try to minimize the bandwidth of this signal since bandwidth is a limited resource. However, spread-spectrum techniques use a transmission bandwidth that is several orders of magnitude greater than the minimum required signal bandwidth. One of the initial reasons for doing this was military applications including guidance and communication systems. These systems were designed using spread spectrum because of its security and resistance to jamming. Asynchronous CDMA has some level of privacy built in because the signal is spread using a pseudo-random code; this code makes the spread-spectrum signals appear random or have noise-like properties. A receiver cannot demodulate this transmission without knowledge of the pseudo-random sequence used to encode the data. CDMA is also resistant to jamming. A jamming signal only has a finite amount of power available to jam the signal. The jammer can either spread its energy over the entire bandwidth of the signal or jam only part of the entire signal.

CDMA can also effectively reject narrow-band interference. Since narrow-band interference affects only a small portion of the spread-spectrum signal, it can easily be removed through notch filtering without much loss of information. Convolution encoding and interleaving can be used to assist in recovering this lost data. CDMA signals are also resistant to multipath fading. Since the spread-spectrum signal occupies a large bandwidth, only a small portion of this will undergo fading due to multipath at any given time. Like the narrow-band interference, this will result in only a small loss of data and can be overcome.

Another reason CDMA is resistant to multipath interference is because the delayed versions of the transmitted pseudo-random codes will have poor correlation with the original pseudo-random code, and will thus appear as another user, which is ignored at the receiver. In other words, as long as the multipath channel induces at least one chip of delay, the multipath signals will arrive at the receiver such that they are shifted in time by at least one chip from the intended signal. The correlation properties of the pseudo-random codes are such that this slight delay causes the multipath to appear uncorrelated with the intended signal, and it is thus ignored.

Some CDMA devices use a rake receiver, which exploits multipath delay components to improve the performance of the system. A rake receiver combines the information from several correlators, each one tuned to a different path delay, producing a stronger version of the signal than a simple receiver with a single correlation tuned to the path delay of the strongest signal.

Frequency reuse is the ability to reuse the same radio channel frequency at other cell sites within a cellular system. In the FDMA and TDMA systems, frequency planning is an important consideration. The frequencies used in different cells must be planned carefully to ensure signals from different cells do not interfere with each other. In a CDMA system, the same frequency can be used in every cell, because channelization is done using the pseudo-random codes. Reusing the same frequency in every cell eliminates the need for frequency planning in a CDMA system; however, planning of the different pseudo-random sequences must be done to ensure that the received signal from one cell does not correlate with the signal from a nearby cell.

Since adjacent cells use the same frequencies, CDMA systems have the ability to perform soft hand-offs. Soft hand-offs allow the mobile telephone to communicate simultaneously with two or more cells. The best signal quality is selected until the hand-off is complete. This is different from hard hand-offs utilized in other cellular systems. In a hard-hand-off situation, as the mobile telephone approaches a hand-off, signal strength may vary abruptly. In contrast, CDMA systems use the soft hand-off, which is undetectable and provides a more reliable and higher-quality signal.

Collaborative CDMA

In a recent study, a novel collaborative multi-user transmission and detection scheme called collaborative CDMA has been investigated for the uplink that exploits the differences between users' fading channel signatures to increase the user capacity well beyond the spreading length in the MAI-limited environment. The authors show that it is possible to achieve this increase at a low complexity and high bit error rate performance in flat fading channels, which is a major research challenge for overloaded CDMA systems. In this approach, instead of using one sequence per user as in conventional CDMA, the authors group a small number of users to share the same spreading sequence and enable group spreading and despreading operations. The new collaborative multi-user receiver consists of two stages: group multi-user detection (MUD) stage to suppress the MAI between the groups and a low-complexity maximum-likelihood detection stage to recover jointly the co-spread users' data using minimal Euclidean-distance measure and users' channel-gain coefficients.

SPATIAL-DIVISION MULTIPLE ACCESS

Spatial division multiple access (SDMA) is a satellite communications mode that optimizes the use of radio spectrum and minimizes system cost by taking advantage of the directional properties of dish antennas. In SDMA, also known as SDM (spatial-division multiplex), satellite dish antennas transmit signals to numerous zones on the earth's surface. The antennas are highly directional, allowing duplicate frequencies to be used for multiple surface zones.

Consider a scenario in which signals must be transmitted simultaneously by one satellite to mobile or portable wireless receivers in 20 different surface zones. In a conventional system, 20 channels and 20 antennas would be necessary to maintain channel separation. In SDMA, there can be far fewer channels than zones. If duplicate-channel zones are sufficiently separated, the 20 signals can be transmitted to earth using four or five channels. The narrow signal beams from the satellite antennas ensure that interference will not occur between zones using the same frequency.

SDMA requires careful choice of zones for each transmitter, and also requires precise antenna alignment. A small error can result in failure of one or more channels, interference among channels, and confusion between surface coverage zones.

References

- Tamburini, F.; Thidé, B.; Mari, E.; Sponselli, A.; Bianchini, A.; Romanato, F. (2012). "Reply to Comment on 'Encoding many channels on the same frequency through radio vorticity: First experimental test'". New Journal of Physics. 14 (11): 118002. Bibcode:2012njph...14k8002t. Doi:10.1088/1367-2630/14/11/118002
- What-is-fdm, network-technologies, computernetworkingnotes: ecomputernotes.com, Retrieved 15 July, 2019
- Guowang Miao; Jens Zander; Ki Won Sung; Ben Slimane (2016). Fundamentals of Mobile Data Networks. Cambridge University Press. ISBN 1107143217
- Fdma-technology, cdma: tutorialspoint.com, Retrieved 16 August, 2019
- She, Alan; Capasso, Federico (17 May 2016). "Parallel Polarization State Generation". Scientific Reports. Nature. 6: 26019. Arxiv:1602.04463. Bibcode:2016natsr...626019S. Doi:10.1038/srep26019. PMC 4869035. PMID 27184813
- Spatial-division-multiple-access, definition: searchnetworking.techtarget.com, Retrieved 17 January, 2019
- María Isabel Gandía Carriedo (August 31, 1998). "ATM: Origins and State of the Art". Universidad Politécnica de Madrid. Archived from the original on June 23, 2006. Retrieved September 23, 2009

5

Telecommunications Networks

The collection of terminal nodes which are linked to enable telecommunication between the terminals is referred to as a telecommunication network. Some of the common telecommunication networks include television network, cellular network, computer network, etc. This chapter has been carefully written to provide an easy understanding of these telecommunication networks.

Telecommunications network is an electronic system of links and switches, and the controls that govern their operation, that allows for data transfer and exchange among multiple users.

When several users of telecommunications media wish to communicate with one another, they must be organized into some form of network. In theory, each user can be given a direct point-to-point link to all the other users in what is known as a fully connected topology (similar to the connections employed in the earliest days of telephony), but in practice this technique is impractical and expensive—especially for a large and dispersed network. Furthermore, the method is inefficient, since most of the links will be idle at any given time. Modern telecommunications networks avoid these issues by establishing a linked network of switches, or nodes, such that each user is connected to one of the nodes. Each link in such a network is called a communications channel. Wire, fibre-optic cable, and radio waves may be used for different communications channels.

Types of Networks

Switched Communications Network

A switched communications network transfers data from source to destination through a series of network nodes. Switching can be done in one of two ways. In a circuit-switched network, a dedicated physical path is established through the network and is held for as long as communication is necessary. An example of this type of network is the traditional (analog) telephone system. A packet-switched network, on the other hand, routes digital data in small pieces called packets, each of which proceeds independently through the network. In a process called store-and-forward, each packet is temporarily stored at each intermediate node, then forwarded when the next link becomes available. In a connection-oriented transmission scheme, each packet takes the same route through the network, and thus all packets usually arrive at the destination in the order in which they were sent. Conversely, each packet may take a different path through the network in a connectionless or datagram scheme. Since datagrams may not arrive at the destination in the order in which they were sent, they are numbered so that they can be properly reassembled. The latter is the method that is used for transmitting data through the Internet.

Broadcast Network

A broadcast network avoids the complex routing procedures of a switched network by ensuring that each node's transmissions are received by all other nodes in the network. Therefore, a broadcast network has only a single communications channel. A wired local area network (LAN), for example, may be set up as a broadcast network, with one user connected to each node and the nodes typically arranged in a bus, ring, or star topology, as shown in the figure. Nodes connected together in a wireless LAN may broadcast via radio or optical links. On a larger scale, many satellite radio systems are broadcast networks, since each Earth station within the system can typically hear all messages relayed by a satellite.

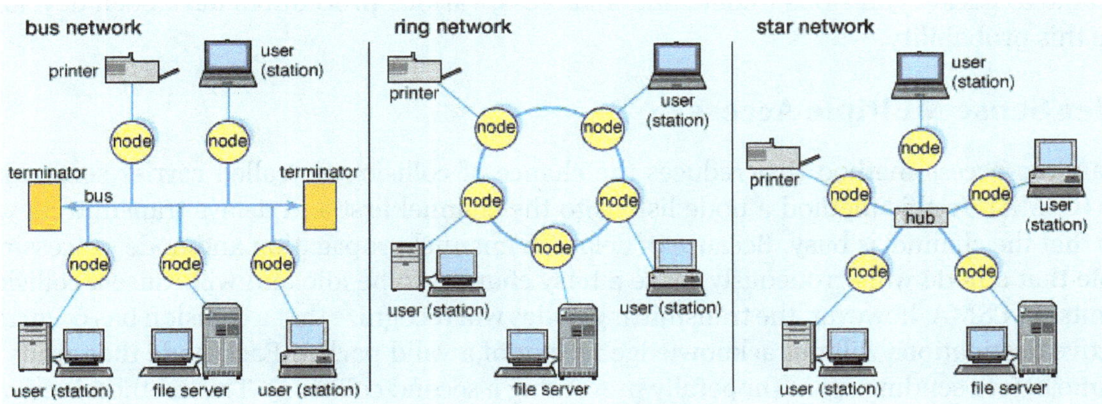

Local area networks (LANs).

Network Access

Since all nodes can hear each transmission in a broadcast network, a procedure must be established for allocating a communications channel to the node or nodes that have packets to transmit and at the same time preventing destructive interference from collisions (simultaneous transmissions). This type of communication, called multiple access, can be established either by scheduling (a technique in which nodes take turns transmitting in an orderly fashion) or by random access to the channel.

Scheduled Access

In a scheduling method known as time-division multiple access (TDMA), a time slot is assigned in turn to each node, which uses the slot if it has something to transmit. If some nodes are much busier than others, then TDMA can be inefficient, since no data are passed during time slots allocated to silent nodes. In this case a reservation system may be implemented, in which there are fewer time slots than nodes and a node reserves a slot only when it is needed for transmission.

A variation of TDMA is the process of polling, in which a central controller asks each node in turn if it requires channel access, and a node transmits a packet or message only in response to its poll.

"Smart" controllers can respond dynamically to nodes that suddenly become very busy by polling them more often for transmissions. A decentralized form of polling is called token passing. In this system a special "token" packet is passed from node to node. Only the node with the token is authorized to transmit; all others are listeners.

Random Access

Scheduled access schemes have several disadvantages, including the large overhead required for the reservation, polling, and token passing processes and the possibility of long idle periods when only a few nodes are transmitting. This can lead to extensive delays in routing information, especially when heavy traffic occurs in different parts of the network at different times—a characteristic of many practical communications networks. Random-access algorithms were designed specifically to give nodes with something to transmit quicker access to the channel. Although the channel is vulnerable to packet collisions under random access, various procedures have been developed to reduce this probability.

Carrier Sense Multiple Access

One random-access method that reduces the chance of collisions is called carrier sense multiple access (CSMA). In this method a node listens to the channel first and delays transmitting when it senses that the channel is busy. Because of delays in channel propagation and node processing, it is possible that a node will erroneously sense a busy channel to be idle and will cause a collision if it transmits. In CSMA, however, the transmitting nodes will recognize that a collision has occurred: the respective destinations will not acknowledge receipt of a valid packet. Each node then waits a random time before sending again (hopefully preventing a second collision). This method is commonly employed in packet networks with radio links, such as the system used by amateur radio operators.

It is important to minimize the time that a communications channel spends in a collision state, since this effectively shuts down the channel. If a node can simultaneously transmit and receive (usually possible on wire and fibre-optic links but not on radio links), then it can stop sending immediately upon detecting the beginning of a collision, thus moving the channel out of the collision state as soon as possible. This process is called carrier sense multiple access with collision detection (CSMA/CD), a feature of the popular wired Ethernet.

Spread-spectrum Multiple Access

Since collisions are so detrimental to network performance, methods have been developed to allow multiple transmissions on a broadcast network without necessarily causing mutual packet destruction. One of the most successful is called spread-spectrum multiple access (SSMA). In SSMA simultaneous transmissions will cause only a slight increase in bit error probability for each user if the channel is not too heavily loaded. Error-free packets can be obtained by using an appropriate control code. Disadvantages of SSMA include wider signal bandwidth and greater equipment cost and complexity compared with conventional CSMA.

Open Systems Interconnection

Different communication requirements necessitate different network solutions, and these different network protocols can create significant problems of compatibility when networks are interconnected with one another. In order to overcome some of these interconnection problems, the open systems interconnection (OSI) was approved in 1983 as an international standard for communications architecture by the International Organization for Standardization (ISO) and the International Telegraph and Telephone Consultative Committee (CCITT). The OSI model, as shown in the figure, consists of seven layers, each of which is selected to perform a well-defined function at a different level of abstraction. The bottom three layers provide for the timely and correct transfer of data, and the top four ensure that arriving data are recognizable and useful. While all seven layers are usually necessary at each user location, only the bottom three are normally employed at a network node, since nodes are concerned only with timely and correct data transfer from point to point.

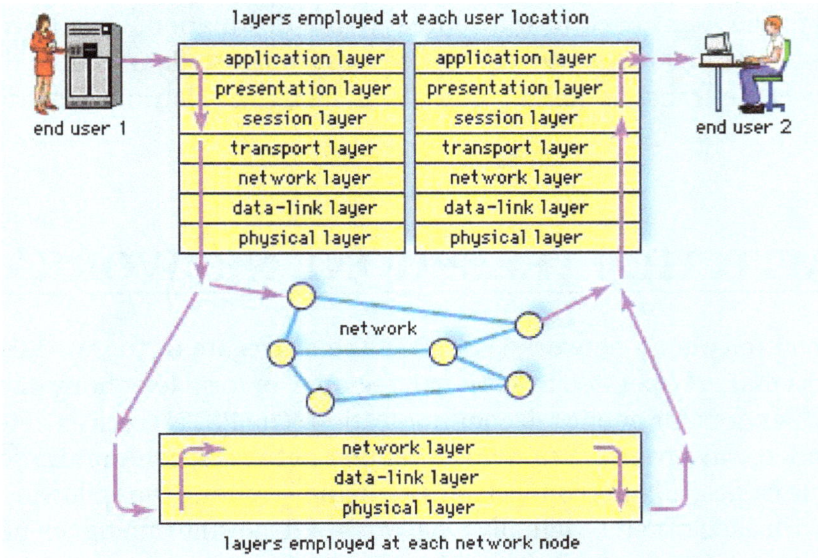

Open systems interconnection.

Data Recognition and Use

The application layer is difficult to generalize, since its content is specific to each user. For example, distributed databases used in the banking and airline industries require several access and security issues to be solved at this level. Network transparency (making the physical distribution of resources irrelevant to the human user) also is handled at this level. The presentation layer, on the other hand, performs functions that are requested sufficiently often that a general solution is warranted. These functions are often placed in a software library that is accessible by several users running different applications. Examples are text conversion, data compression, and data encryption.

User interface with the network is performed by the session layer, which handles the process of connecting to another computer, verifying user authenticity, and establishing a reliable communication process. This layer also ensures that files which can be altered by several network users are kept in order. Data from the session layer are accepted by the transport layer, which separates the data stream into smaller units, if necessary, and ensures that all arrive correctly at the destination. If fast throughput is needed, the transport layer may establish several simultaneous paths in the network and send different parts of the data over each path. Conversely, if low cost is a requirement, then the layer may time-multiplex several users' data over one path through the network. Flow control is also regulated at this level, ensuring that data from a fast source will not overrun a slow destination.

Data Transfer

The network layer breaks data into packets and determines how the packets are routed within the network, which nodes (if any) will check packets for errors along the route, and whether congestion control is needed in a heavily loaded network. The data-link layer transforms a raw communications channel into a line that appears essentially free of transmission errors to the network layer. This is done by breaking data up into data frames, transmitting them sequentially, and processing acknowledgment frames sent back to the source by the destination. This layer also establishes frame boundaries and implements recovery procedures from lost, damaged, or duplicated frames. The physical layer is the transmission medium itself, along with various electric and mechanical specifications.

PUBLIC SWITCHED TELEPHONE NETWORK

The public switched telephone network (PSTN) is the aggregate of the world's circuit-switched telephone networks that are operated by national, regional, or local telephony operators, providing infrastructure and services for public telecommunication. The PSTN consists of telephone lines, fiber optic cables, microwave transmission links, cellular networks, communications satellites, and undersea telephone cables, all interconnected by switching centers, thus allowing most telephones to communicate with each other. Originally a network of fixed-line analog telephone systems, the PSTN is now almost entirely digital in its core network and includes mobile and other networks, as well as fixed telephones.

The technical operation of the PSTN adheres to the standards created by the ITU-T. These standards allow different networks in different countries to interconnect seamlessly. The E.163 and E.164 standards provide a single global address space for telephone numbers. The combination of the interconnected networks and the single numbering plan allow telephones around the world to dial each other.

Operators

The task of building the networks and selling services to customers fell to the network operators. The first company to be incorporated to provide PSTN services was the Bell Telephone Company in the United States.

In some countries, however, the job of providing telephone networks fell to government as the investment required was very large and the provision of telephone service was increasingly becoming an essential public utility. For example, the General Post Office in the United Kingdom brought together a number of private companies to form a single nationalized company. In more recent decades, these state monopolies were broken up or sold off through privatization.

Regulation

In most countries, the central has a regulator dedicated to monitoring the provision of PSTN services in that country. Their tasks may be for example to ensure that end customers are not overcharged for services where monopolies may exist. These regulatory agencies may also regulate the prices charged between the operators to carry each other's traffic.

Technology

Network Topology

The PSTN network architecture had to evolve over the years to support increasing numbers of subscribers, calls, connections to other countries, direct dialing and so on. The model developed by the United States and Canada was adopted by other nations, with adaptations for local markets.

The original concept was that the telephone exchanges are arranged into hierarchies, so that if a call cannot be handled in a local cluster, it is passed to one higher up for onward routing. This reduced the number of connecting trunks required between operators over long distances and also kept local traffic separate.

However, in modern networks the cost of transmission and equipment is lower and, although hierarchies still exist, they are much flatter, with perhaps only two layers.

Digital Channels

Most automated telephone exchanges use digital switching rather than mechanical or analog switching. The trunks connecting the exchanges are also digital, called circuits or channels. However analog two-wire circuits are still used to connect the last mile from the exchange to the telephone in the home (also called the local loop). To carry a typical phone call from a calling party to a called party, the analog audio signal is digitized at an 8 kHz sample rate with 8-bit resolution using a special type of nonlinear pulse code modulation known as G.711. The call is then transmitted from one end to another via telephone exchanges. The call is switched using a call set up protocol (usually ISUP) between the telephone exchanges under an overall routing strategy.

The call is carried over the PSTN using a 64 kbit/s channel, originally designed by Bell Labs. The name given to this channel is Digital Signal 0 (DS0). The DS0 circuit is the basic granularity of circuit switching in a telephone exchange. A DS0 is also known as a timeslot because DS0s are aggregated in time-division multiplexing (TDM) equipment to form higher capacity communication links.

A Digital Signal 1 (DS1) circuit carries 24 DS0s on a North American or Japanese T-carrier (T1) line, or 32 DS0s (30 for calls plus two for framing and signaling) on an E-carrier (E1) line used in most other countries. In modern networks, the multiplexing function is moved as close to the end user as possible, usually into cabinets at the roadside in residential areas, or into large business premises.

These aggregated circuits are conveyed from the initial multiplexer to the exchange over a set of equipment collectively known as the access network. The access network and inter-exchange transport use synchronous optical transmission, for example, SONET and Synchronous Digital Hierarchy (SDH) technologies, although some parts still use the older PDH technology.

Within the access network, there are a number of reference points defined. Most of these are of interest mainly to ISDN but one – the V reference point – is of more general interest. This is the reference point between a primary multiplexer and an exchange. The protocols at this reference point were standardized in ETSI areas as the V5 interface.

TELEVISION NETWORK

A television network is a telecommunications network for distribution of television program content, whereby a central operation provides programming to many television stations or pay television providers. Until the mid-1980s, television programming in most countries of the world was dominated by a small number of terrestrial networks. Many early television networks (such as the BBC, NBC or CBC) evolved from earlier radio networks.

In countries where most networks broadcast identical, centrally originated content to all of their stations and where most individual television transmitters therefore operate only as large "repeater stations", the terms "television network", "television channel" (a numeric identifier or radio frequency) and "television station" have become mostly interchangeable in everyday language, with professionals in television-related occupations continuing to make a differentiation between them. Within the industry, a tiering is sometimes created among groups of networks based on whether their programming is simultaneously originated from a central point, and whether the network master control has the technical and administrative capability to take over the programming of their affiliates in real-time when it deems this necessary – the most common example being during national breaking news events.

In North America in particular, many television networks available via cable and satellite television are branded as "channels" because they are somewhat different from traditional networks in the sense defined above, as they are singular operations – they have no affiliates or component stations, but instead are distributed to the public via cable or direct-broadcast satellite providers. Such networks are commonly referred to by terms such as "specialty channels" in Canada or "cable networks" in the U.S.

A network may or may not produce all of its own programming. If not, production companies (such as Warner Bros. Television, Universal Television, Sony Pictures Television and TriStar Television) can distribute their content to the various networks, and it is common that a certain production firm may have programs that air on two or more rival networks. Similarly, some networks may import television programs from other countries, or use archived programming to help complement their schedules.

Some stations have the capability to interrupt the network through the local insertion of television commercials, station identifications and emergency alerts. Others completely break away from the network for their own programming, a method known as regional variation. This is common where small networks are members of larger networks. The majority of commercial television stations are self-owned, even though a variety of these instances are the property of an owned-and-operated television network. The commercial television stations can also be linked with a noncommercial educational broadcasting agency. It is also important to note that some countries have launched national television networks, so that individual television stations can act as common repeaters of nationwide programs.

On the other hand, television networks also undergo the impending experience of major changes related to cultural varieties. The emergence of cable television has made available in major media markets, programs such as those aimed at American bi-cultural Latinos. Such a diverse captive audience presents an occasion for the networks and affiliates to advertise the best programming that needs to be aired.

This is explained by author Tim P. Vos in his abstract A Cultural Explanation of Early Broadcast, where he determines targeted group/non-targeted group representations as well as the cultural specificity employed in the television network entity. Vos notes that policymakers did not expressly intend to create a broadcast order dominated by commercial networks. In fact, legislative attempts were made to limit the network's preferred position.

As to individual stations, modern network operations centers usually use broadcast automation to handle most tasks. These systems are not only used for programming and for video server playout, but use exact atomic time from Global Positioning Systems or other sources to maintain perfect synchronization with upstream and downstream systems, so that programming appears seamless to viewers.

Global

A major international television network is the British Broadcasting Corporation (BBC), which is perhaps most well known for its news agency BBC News. Owned by the Crown, the BBC operates primarily in the United Kingdom. It is funded by the television licence paid by British residents that watch terrestrial television and as a result, no commercial advertising appears on its networks. Outside the UK, advertising is broadcast because the licence fee only applies to the BBC's British operations. 23,000 people worldwide are employed by the BBC and its subsidiary, BBC Studios.

United States

Television in the United States had long been dominated by the Big Three television networks, the American Broadcasting Company (ABC), CBS (formerly the Columbia Broadcasting System) and the National Broadcasting Company (NBC); however the Fox Broadcasting Company (Fox), which launched in October 1986, has gained prominence and is now considered part of the "Big Four." The Big Three provide a significant amount of programs to each of their affiliates, including newscasts, prime time, daytime and sports programming, but still reserve periods during each day where their affiliate can air local programming, such as local news or syndicated programs. Since the creation of Fox, the number of American television networks has increased, though the amount of programming they provide is often much less: for example, The CW Television Network only provides twelve hours of primetime programming each week (along with six hours on Saturdays and five hours a week during the daytime), leaving its affiliates to fill time periods where network programs are not broadcast with a large amount of syndicated programming. Other networks are dedicated to specialized programming, such as religious content or programs presented in languages other than English, particularly Spanish.

The largest television network in the United States, however, is the Public Broadcasting Service (PBS), a non-profit, publicly owned, non-commercial educational service. In comparison to the commercial television networks, there is no central unified arm of broadcast programming, meaning that each PBS member station has a significant amount of freedom to schedule television shows as they consent to. Some public television outlets, such as PBS, carry separate digital subchannel networks through their member stations (for example, Georgia Public Broadcasting; in fact, some programs airing on PBS were branded on other channels as coming from GPB Kids and PBS World).

This works as each network sends its signal to many local affiliated television stations across the country. These local stations then carry the "network feed," which can be viewed by millions of households across the country. In such cases, the signal is sent to as many as 200+ stations or as little as just a dozen or fewer stations, depending on the size of the network.

With the adoption of digital television, television networks have also been created specifically for distribution on the digital subchannels of television stations (including networks focusing on classic television series and films operated by companies like Weigel Broadcasting (owners of Movies and Me-TV) and Tribune Broadcasting (owners of this TV and Antenna TV), along with networks focusing on music, sports and other niche programming).

Cable and satellite providers pay the networks a certain rate per subscriber (the highest charge being for ESPN, in which cable and satellite providers pay a rate of more than $5.00 per subscriber to ESPN). The providers also handle the sale of advertising inserted at the local level during national programming, in which case the broadcaster and the cable/satellite provider may share revenue. Networks that maintain a home shopping or infomercial format may instead pay the station or cable/satellite provider, in a brokered carriage deal. This is especially common with low-power television stations, and in recent years, even more so for stations that used this revenue stream to finance their conversion to digital broadcasts, which in turn provides them with several additional channels to transmit different programming sources.

Regulation

FCC regulations in the United States restricted the number of television stations that could be owned by any one network, company or individual. This led to a system where most local television stations were independently owned, but received programming from the network through a franchising contract, except in a few major cities that had owned-and-operated stations (O&O) of a network and independent stations. In the early days of television, when there were often only one or two stations broadcasting in a given market, the stations were usually affiliated with multiple networks and were able to choose which programs would air. Eventually, as more stations were licensed, it became common for each station to be exclusively affiliated with only one network and carry all of the "prime-time" programs that the network offered. Local stations occasionally break from regularly scheduled network programming however, especially when a breaking news or severe weather situation occurs in the viewing area. Moreover, when stations return to network programming from commercial breaks, station identifications are displayed in the first few seconds before switching to the network's logo.

Canada

A number of different definitions of "network" are used by government agencies, industry, and the general public. Under the Broadcasting Act, a network is defined as "any operation where control over all or any part of the programs or program schedules of one or more broadcasting undertakings is delegated to another undertaking or person," and must be licensed by the Canadian Radio-television and Telecommunications Commission (CRTC).

Only three national over-the-air television networks are currently licensed by the CRTC: government-owned CBC Television (English) and Ici Radio-Canada Télé (French), French-language private network TVA, and a network focused on Canada's indigenous peoples. A third French-language service, V, is licensed as a provincial network within Quebec, but is not licensed or locally distributed (outside of carriage on the digital tiers of pay television providers) on a national basis.

Currently, licensed national or provincial networks must be carried by all cable providers (in the country or province, respectively) with a service area above a certain population threshold, as well as all satellite providers. However, they are no longer necessarily expected to achieve over-the-air coverage in all areas (APTN, for example, only has terrestrial coverage in parts of northern Canada).

In addition to these licensed networks, the two main private English-language over-the-air services, CTV and Global, are also generally considered to be "networks" by virtue of their national coverage, although they are not officially licensed as such. CTV was previously a licensed network, but relinquished this licence in 2001 after acquiring most of its affiliates, making operating a network licence essentially redundant.

Smaller groups of stations with common branding are often categorized by industry watchers as television systems, although the public and the broadcasters themselves will often refer to them as "networks" regardless. Some of these systems, such as CTV Two and the now-defunct E!, essentially operate as mini-networks, but have reduced geographical coverage. Others, such as Omni Television or the Crossroads Television System, have similar branding and a common programming focus, but schedules may vary significantly from one station to the next. Citytv originally began operating as a television system in 2002 when CKVU-TV in Vancouver started to carry programs originating from CITY-TV in Toronto and adopted that station's "Citytv" branding, but gradually became a network by virtue of national coverage through expansions into other markets west of Atlantic Canada between 2005 and 2013.

Most local television stations in Canada are now owned and operated directly by their network, with only a small number of stations still operating as affiliates.

Europe, Asia, Africa and South America

Most television services outside North America are national networks established by a combination of publicly funded broadcasters and commercial broadcasters. Most nations established television networks in a similar way: the first television service in each country was operated by a public broadcaster, often funded by a television licensing fee, and most of them later established a second or even third station providing a greater variety of content. Commercial television services also became available when private companies applied for television broadcasting licenses. Often, each new network would be identified with their channel number, so that individual stations would often be numbered "One," "Two," "Three," and so forth.

United Kingdom

The first television network in the United Kingdom was operated by the British Broadcasting Corporation (BBC). On 2 November 1936 the BBC opened the world's first regular high-definition television service, from a 405 lines transmitter at Alexandra Palace. The BBC remained dominant until eventually on 22 September 1955, commercial broadcasting was established in order to create a second television network. Rather than creating a single network with local stations owned and operated by a single company (as is the case with the BBC), each local area had a separate television station that was independently owned and operated, although most of these stations shared a number of programmes, particularly during peak evening viewing hours. These stations formed the ITV network.

When the advent of UHF broadcasting allowed a greater number of television stations to broadcast, the BBC launched a second network, BBC Two (with the original service being renamed BBC One). A fourth national commercial service was launched, Channel 4, although Wales instead introduced a Welsh-language service, S4C. These were later followed by the launch of a fifth network, Channel 5. Since the introduction of digital television, the BBC, ITV, Channel 4 and Channel 5 each introduced a number of digital-only networks. Sky operates a large number of networks including Sky One, Sky Living and Sky Atlantic; as does UKTV, which operates networks like Dave, Gold, W and Yesterday.

Sweden

Sweden had only one television network until the early-1990s: the public broadcaster Sveriges Television (SVT). Commercial companies such as Modern Times Group, TV4, Viasat, and SBS Discovery have established TV networks since the 1980s although they initially aired exclusively on satellite. In 1991, TV4 became Sweden's first commercial television network to air terrestrially. Most television programming in Sweden is centralised except for local news updates that air on SVT2 and TV4.

Netherlands

Until 1989, Netherlands Public Broadcasting was the only television network in the Netherlands, with three stations, Nederland 1, Nederland 2 and Nederland 3. Rather than having a single production arm, there are a number of public broadcasting organizations that create programming for each of the three stations, each working relatively independently. Commercial broadcasting in the Netherlands is currently operated by two networks, RTL Nederland and SBS Broadcasting, which together broadcast seven commercial stations.

Russia

Soviet Era

The first television network in the Soviet Union launched on 7 July 1938 when Petersburg – Channel 5 of Leningrad Television became a unionwide network. The second television network in the Soviet Union launched on 22 March 1951 when Channel One of USSR Central Television became a unionwide network. Until 1989, there were six television networks, all owned by the USSR

Gosteleradio. This changed during Mikhail Gorbachev's Perestroika program, when the first independent television network, 2×2, was launched.

1990s

Following the breakup of the Soviet Union, USSR Gosteleradio ceased to exist as well as its six networks. Only Channel One had a smooth transition and survived as a network, becoming Ostankino Channel One. The other five networks were operated by Ground Zero. This free airwave space allowed many private television networks like NTV and TV-6 to launch in the mid-1990s.

2000s

The 2000s were marked by the increased state intervention in Russian television. On April 14, 2001 NTV experienced management changes following the expulsion of former oligarch and NTV founder Vladimir Gusinsky. As a result, most of the prominent reporters featured on NTV left the network. Later on January 22, 2002, the second largest private television network TV-6, where the former NTV staff took refuge, was shut down allegedly because of its editorial policy. Five months later on June 1, TVS was launched, mostly employing NTV/TV-6 staff, only to cease operations the following year. Since then, the four largest television networks (Channel One, Russia 1, NTV and Russia 2) have been state-owned.

Still, the 2000s saw a rise of several independent television networks such as REN (its coverage increased vastly allowing it to become a federal network), Petersburg – Channel Five (overall the same), the relaunched 2×2. The Russian television market is mainly shared today by five major companies: Channel One, Russia 1, NTV, TNT and CTC.

Brazil

The major commercial television network in Brazil is Rede Globo, which was founded in 1965. It grew to become the largest and most successful media conglomerate in the country, having a dominating presence in various forms of media including television, radio, print (newspapers and magazines) and the Internet. Other networks include Rede Bandeirantes, RecordTV, SBT, RedeTV!, and TV Cultura.

Australia

Australia has two national public networks, ABC Television and SBS. The ABC operates eight stations as part of its main network ABC, one for each state and territory, as well as three digital-only networks, ABC Kids/ABC Comedy, ABC Me and ABC News. SBS currently operates four stations, SBS, SBS Viceland, SBS Food and NITV.

The first commercial networks in Australia involved commercial stations that shared programming in Sydney, Melbourne, Brisbane, Adelaide and later Perth, with each network forming networks based on their allocated channel numbers: TCN-9 in Sydney, GTV-9 in Melbourne, QTQ-9 in Brisbane, NWS-9 in Adelaide and STW-9 in Perth together formed the Nine Network; while their equivalents on VHF channels 7 and 10 respectively formed the Seven Network and Network

Ten. Until 1989, areas outside these main cities had access to only a single commercial station, and these rural stations often formed small networks such as Prime Television. Beginning in 1989, however, television markets in rural areas began to aggregate, allowing these rural networks to broadcast over a larger area, often an entire state, and become full-time affiliates to one specific metropolitan network.

As well as these Free-to-air channels, there are other's on Australia's Pay television network Foxtel.

New Zealand

New Zealand has one public network, Television New Zealand (TVNZ), which consists of two main networks: TVNZ 1 is the network's flagship network which carries news, current affairs and sports programming as well as the majority of the locally produced shows broadcast by TVNZ and imported shows. TVNZ's second network, TV2, airs mostly imported shows with some locally produced programs such as *Shortland Street*. TVNZ also operates a network exclusive to pay television services, TVNZ Heartland, available on providers such as Sky. TVNZ previously operated a non-commercial public service network, TVNZ 7, which ceased operations in June 2012 and was replaced by the timeshift channel TV One Plus 1. The network operated by Television New Zealand has progressed from operating as four distinct local stations within the four main centers in the 1960s, to having the majority of the content produced from TVNZ's Auckland studios at present.

New Zealand also has several privately owned television networks with the largest being operated by MediaWorks. MediaWorks' flagship network is TV3, which competes directly with both TVNZ broadcast networks. MediaWorks also operates a second network, FOUR, which airs mostly imported programmes with children's shows airing in the daytime and shows targeted at teenagers and adult between 15 and 39 years of age during prime time. MediaWorks also operates a timeshift network, TV3 + 1, and a 24-hour music network, C4.

All television networks in New Zealand air the same programming across the entire country with the only regional deviations being for local advertising; a regional news service existed in the 1980s, carrying a regional news programme from TVNZ's studios in New Zealand's four largest cities, Auckland, Wellington, Christchurch and Dunedin.

In the 1960s, the service operated at the time by the New Zealand Broadcasting Corporation was four separate television stations – AKTV2 in Auckland, WNTV1 in Wellington, CHTV3 in Christchurch and DNTV2 in Dunedin – which each ran their own newscast and produced some in-house programmes, with other shows being shared between the stations. Programmes and news footage were distributed via mail, with a programme airing in one region being mailed to another region for broadcast the following week. A network was finally established in 1969, with the same programmes being relayed to all regions simultaneously. From the 1970s to the 1990s, locally produced programmes that aired on TV One and TV2 were produced out of one of the four main studios, with TVNZ's network hub based in Wellington. Today, most locally produced programmes that are aired by both TVNZ and other networks are not actually produced in-house, instead they are often produced by a third party company (for example, the TV2 programme *Shortland Street* is produced by South Pacific Pictures). The networks produce their own news and current affairs programs, with most of the content filmed in Auckland.

New Zealand also operates several regional television stations, which are only available in individual markets. The regional stations will typically air a local news programme, produce some shows in-house and cover local sports events; the majority of programming on the regional stations will be imported from various sources.

Philippines

In the Philippines, in practice, the terms "network," "station" and "channel" are used interchangeably as programming lineups are mostly centrally planned from the networks' main offices, and since provincial/regional stations usually just relay the broadcast from their parent network's flagship station (usually based in the Mega Manila area). As such, networks made up of VHF stations are sometimes informally referred to by their over-the-air channel number in the Mega Manila area (for example, Channel 2 or *Dos* for ABS-CBN, Channel 5 or *Singko* for TV5, and Channel 7 or *Siyete* for GMA Network), while some incorporate their channel numbers in the network's name (for example, TV5, Studio 23 and Net 25, which respectively broadcast on VHF channel 5, and UHF channels 23 and 25).

Unlike the United States, where networks receive programmes produced by various production companies, the two largest networks in the Philippines produce all of their prime time programmes except for Asianovelas. Other networks adopt block-time programming, which utilizes programming arrangements similar to the relationship between a U.S. network and station.

CELLULAR NETWORK

A cellular network or mobile network is a communication network where the last link is wireless. The network is distributed over land areas called "cells", each served by at least one fixed-location transceiver, but more normally, three cell sites or base transceiver stations. These base stations provide the cell with the network coverage which can be used for transmission of voice, data, and other types of content. A cell typically uses a different set of frequencies from neighbouring cells, to avoid interference and provide guaranteed service quality within each cell.

When joined together, these cells provide radio coverage over a wide geographic area. This enables numerous portable transceivers (e.g., mobile phones, tablets and laptops equipped with mobile broadband modems, pagers, etc.) to communicate with each other and with fixed transceivers and telephones anywhere in the network, via base stations, even if some of the transceivers are moving through more than one cell during transmission.

Cellular networks offer a number of desirable features:

- More capacity than a single large transmitter, since the same frequency can be used for multiple links as long as they are in different cells.
- Mobile devices use less power than with a single transmitter or satellite since the cell towers are closer.
- Larger coverage area than a single terrestrial transmitter, since additional cell towers can be added indefinitely and are not limited by the horizon.

Major telecommunications providers have deployed voice and data cellular networks over most of the inhabited land area of Earth. This allows mobile phones and mobile computing devices to be connected to the public switched telephone network and public Internet. Private cellular networks can be used for research or for large organizations and fleets, such as dispatch for local public safety agencies or a taxicab company.

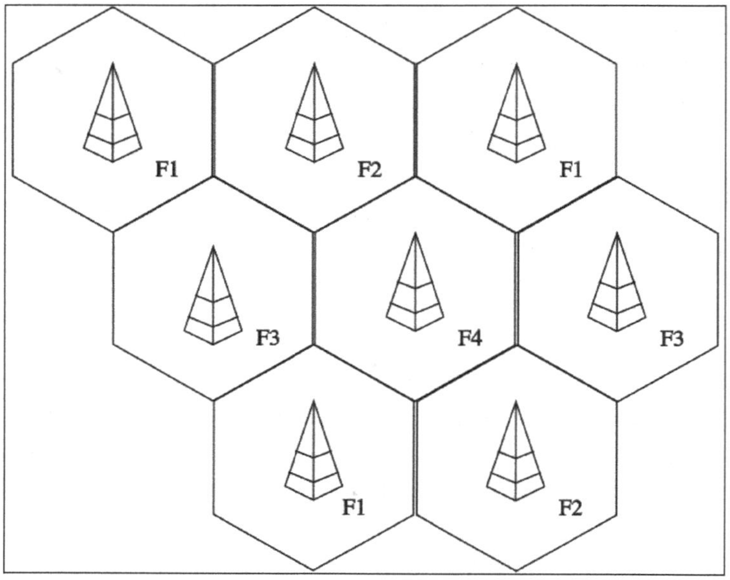

Example of frequency reuse factor or pattern 1/4.

In a cellular radio system, a land area to be supplied with radio service is divided into cells in a pattern dependent on terrain and reception characteristics. These cell patterns roughly take the form of regular shapes, such as hexagons, squares, or circles although hexagonal cells are conventional. Each of these cells is assigned with multiple frequencies ($f_1 - f_6$) which have corresponding radio base stations. The group of frequencies can be reused in other cells, provided that the same frequencies are not reused in adjacent cells, which would cause co-channel interference.

The increased capacity in a cellular network, compared with a network with a single transmitter, comes from the mobile communication switching system developed by Amos Joel of Bell Labs that permitted multiple callers in a given area to use the same frequency by switching calls to the nearest available cellular tower having that frequency available. This strategy is viable because a given radio frequency can be reused in a different area for an unrelated transmission. In contrast, a single transmitter can only handle one transmission for a given frequency. Inevitably, there is some level of interference from the signal from the other cells which use the same frequency. Consequently, there must be at least one cell gap between cells which reuse the same frequency in a standard FDMA system.

Consider the case of a taxi company, where each radio has a manually operated channel selector knob to tune to different frequencies. As drivers move around, they change from channel to channel. The drivers are aware of which frequency approximately covers some area. When they do not receive a signal from the transmitter, they try other channels until finding one that works. The taxi drivers only speak one at a time when invited by the base station operator. This is a form of time-division multiple access (TDMA).

The first commercial cellular network, the 1G generation, was launched in Japan by Nippon Telegraph and Telephone (NTT) in 1979, initially in the metropolitan area of Tokyo. Within five years, the NTT network had been expanded to cover the whole population of Japan and became the first nationwide 1G network.

Cell Signal Encoding

To distinguish signals from several different transmitters, time-division multiple access (TDMA), frequency-division multiple access (FDMA), code-division multiple access (CDMA), and orthogonal frequency-division multiple access (OFDMA) were developed.

With TDMA, the transmitting and receiving time slots used by different users in each cell are different from each other.

With FDMA, the transmitting and receiving frequencies used by different users in each cell are different from each other. In a simple taxi system, the taxi driver manually tuned to a frequency of a chosen cell to obtain a strong signal and to avoid interference from signals from other cells.

The principle of CDMA is more complex, but achieves the same result; the distributed transceivers can select one cell and listen to it.

Other available methods of multiplexing such as polarization-division multiple access (PDMA) cannot be used to separate signals from one cell to the next since the effects of both vary with position and this would make signal separation practically impossible. TDMA is used in combination with either FDMA or CDMA in a number of systems to give multiple channels within the coverage area of a single cell.

Frequency Reuse

The key characteristic of a cellular network is the ability to re-use frequencies to increase both coverage and capacity. As described above, adjacent cells must use different frequencies, however, there is no problem with two cells sufficiently far apart operating on the same frequency, provided the masts and cellular network users' equipment do not transmit with too much power.

The elements that determine frequency reuse are the reuse distance and the reuse factor. The reuse distance, D is calculated as:

$$D = R\sqrt{3N},$$

where R is the cell radius and N is the number of cells per cluster. Cells may vary in radius from 1 to 30 kilometres (0.62 to 18.64 mi). The boundaries of the cells can also overlap between adjacent cells and large cells can be divided into smaller cells.

The frequency reuse factor is the rate at which the same frequency can be used in the network. It is $1/K$ (or K according to some books) where K is the number of cells which cannot use the same frequencies for transmission. Common values for the frequency reuse factor are 1/3, 1/4, 1/7, 1/9 and 1/12 (or 3, 4, 7, 9 and 12 depending on notation).

In case of *N* sector antennas on the same base station site, each with different direction, the base station site can serve N different sectors. *N* is typically 3. A reuse pattern of *N/K* denotes a further division in frequency among *N* sector antennas per site. Some current and historical reuse patterns are 3/7 (North American AMPS), 6/4 (Motorola NAMPS), and 3/4 (GSM).

If the total available bandwidth is *B*, each cell can only use a number of frequency channels corresponding to a bandwidth of *B/K*, and each sector can use a bandwidth of *B/NK*.

Code-division multiple access-based systems use a wider frequency band to achieve the same rate of transmission as FDMA, but this is compensated for by the ability to use a frequency reuse factor of 1, for example using a reuse pattern of 1/1. In other words, adjacent base station sites use the same frequencies, and the different base stations and users are separated by codes rather than frequencies. While *N* is shown as 1 in this example, that does not mean the CDMA cell has only one sector, but rather that the entire cell bandwidth is also available to each sector individually.

Depending on the size of the city, a taxi system may not have any frequency-reuse in its own city, but certainly, in other nearby cities, the same frequency can be used. In a large city, on the other hand, frequency-reuse could certainly be in use.

Recently also orthogonal frequency-division multiple access based systems such as LTE are being deployed with a frequency reuse of 1. Since such systems do not spread the signal across the frequency band, inter-cell radio resource management is important to coordinate resource allocation between different cell sites and to limit the inter-cell interference. There are various means of Inter-Cell Interference Coordination (ICIC) already defined in the standard. Coordinated scheduling, multi-site MIMO or multi-site beamforming are other examples for inter-cell radio resource management that might be standardized in the future.

Directional Antennas

Cell towers frequently use a directional signal to improve reception in higher-traffic areas. In the United States, the Federal Communications Commission (FCC) limits omnidirectional cell tower signals to 100 watts of power. If the tower has directional antennas, the FCC allows the cell operator to broadcast up to 500 watts of effective radiated power (ERP).

Although the original cell towers created an even, omnidirectional signal, were at the centers of the cells and were omnidirectional, a cellular map can be redrawn with the cellular telephone towers located at the corners of the hexagons where three cells converge. Each tower has three sets of directional antennas aimed in three different directions with 120 degrees for each cell (totaling 360 degrees) and receiving/transmitting into three different cells at different frequencies. This provides a minimum of three channels, and three towers for each cell and greatly increases the chances of receiving a usable signal from at least one direction.

The numbers in the illustration are channel numbers, which repeat every 3 cells. Large cells can be subdivided into smaller cells for high volume areas.

Cell phone companies also use this directional signal to improve reception along highways and inside buildings like stadiums and arenas.

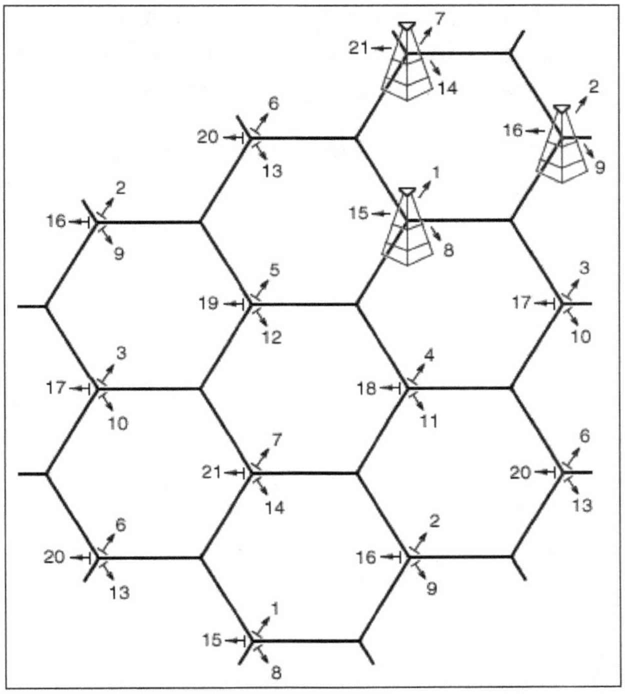

Cellular telephone frequency reuse pattern.

Broadcast Messages and Paging

Practically every cellular system has some kind of broadcast mechanism. This can be used directly for distributing information to multiple mobiles. Commonly, for example in mobile telephony systems, the most important use of broadcast information is to set up channels for one-to-one communication between the mobile transceiver and the base station. This is called paging. The three different paging procedures generally adopted are sequential, parallel and selective paging.

The details of the process of paging vary somewhat from network to network, but normally we know a limited number of cells where the phone is located (this group of cells is called a Location Area in the GSM or UMTS system, or Routing Area if a data packet session is involved; in LTE, cells are grouped into Tracking Areas). Paging takes place by sending the broadcast message to all of those cells. Paging messages can be used for information transfer. This happens in pagers, in CDMA systems for sending SMS messages, and in the UMTS system where it allows for low downlink latency in packet-based connections.

Movement from Cell to Cell and Handing Over

In a primitive taxi system, when the taxi moved away from a first tower and closer to a second tower, the taxi driver manually switched from one frequency to another as needed. If communication was interrupted due to a loss of a signal, the taxi driver asked the base station operator to repeat the message on a different frequency.

In a cellular system, as the distributed mobile transceivers move from cell to cell during an ongoing continuous communication, switching from one cell frequency to a different cell frequency is done electronically without interruption and without a base station operator or manual switching.

This is called the handover or handoff. Typically, a new channel is automatically selected for the mobile unit on the new base station which will serve it. The mobile unit then automatically switches from the current channel to the new channel and communication continues.

The exact details of the mobile system's move from one base station to the other vary considerably from system to system.

Mobile Phone Network

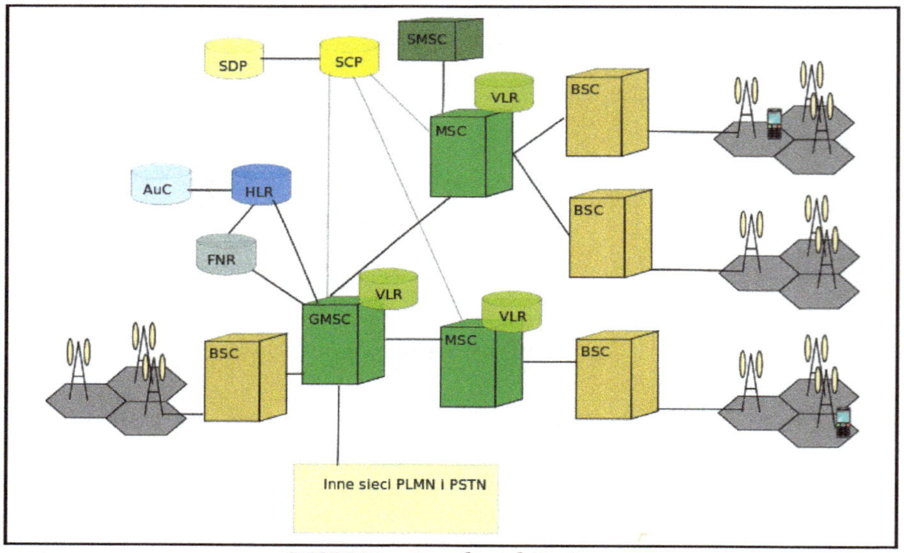

WCDMA network architecture.

The most common example of a cellular network is a mobile phone (cell phone) network. A mobile phone is a portable telephone which receives or makes calls through a cell site (base station) or transmitting tower. Radio waves are used to transfer signals to and from the cell phone.

Modern mobile phone networks use cells because radio frequencies are a limited, shared resource. Cell-sites and handsets change frequency under computer control and use low power transmitters so that the usually limited number of radio frequencies can be simultaneously used by many callers with less interference.

A cellular network is used by the mobile phone operator to achieve both coverage and capacity for their subscribers. Large geographic areas are split into smaller cells to avoid line-of-sight signal loss and to support a large number of active phones in that area. All of the cell sites are connected to telephone exchanges (or switches), which in turn connect to the public telephone network.

In cities, each cell site may have a range of up to approximately 1/2 mile (0.80 km), while in rural areas, the range could be as much as 5 miles (8.0 km). It is possible that in clear open areas, a user may receive signals from a cell site 25 miles (40 km) away.

Since almost all mobile phones use cellular technology, including GSM, CDMA, and AMPS (analog), the term "cell phone" is in some regions, notably the US, used interchangeably with "mobile phone". However, satellite phones are mobile phones that do not communicate directly with a ground-based cellular tower but may do so indirectly by way of a satellite.

There are a number of different digital cellular technologies, including: Global System for Mobile Communications (GSM), General Packet Radio Service (GPRS), cdmaOne, CDMA2000, Evolution-Data Optimized (EV-DO), Enhanced Data Rates for GSM Evolution (EDGE), Universal Mobile Telecommunications System (UMTS), Digital Enhanced Cordless Telecommunications (DECT), Digital AMPS (IS-136/TDMA), and Integrated Digital Enhanced Network (iDEN). The transition from existing analog to the digital standard followed a very different path in Europe and the US. As a consequence, multiple digital standards surfaced in the US, while Europe and many countries converged towards the GSM standard.

Structure of the Mobile Phone Cellular Network

A simple view of the cellular mobile-radio network consists of the following:

- A network of radio base stations forming the base station subsystem.
- The core circuit switched network for handling voice calls and text.
- A packet switched network for handling mobile data.
- The public switched telephone network to connect subscribers to the wider telephony network.

This network is the foundation of the GSM system network. There are many functions that are performed by this network in order to make sure customers get the desired service including mobility management, registration, call set-up, and handover.

Any phone connects to the network via an RBS (Radio Base Station) at a corner of the corresponding cell which in turn connects to the Mobile switching center (MSC). The MSC provides a connection to the public switched telephone network (PSTN). The link from a phone to the RBS is called an *uplink* while the other way is termed *downlink*.

Radio channels effectively use the transmission medium through the use of the following multiplexing and access schemes: frequency division multiple access (FDMA), time division multiple access (TDMA), code division multiple access (CDMA), and space division multiple access (SDMA).

Small Cells

Small cells, which have a smaller coverage area than base stations, are categorised as follows:

- Microcell, less than 2 kilometres.
- Picocell, less than 200 metres.
- Femtocell, around 10 metres.

Cellular Handover in Mobile Phone Networks

As the phone user moves from one cell area to another cell while a call is in progress, the mobile station will search for a new channel to attach to in order not to drop the call. Once a new channel is found, the network will command the mobile unit to switch to the new channel and at the same time switch the call onto the new channel.

With CDMA, multiple CDMA handsets share a specific radio channel. The signals are separated by using a pseudonoise code (PN code) that is specific to each phone. As the user moves from one cell to another, the handset sets up radio links with multiple cell sites (or sectors of the same site) simultaneously. This is known as "soft handoff" because, unlike with traditional cellular technology, there is no one defined point where the phone switches to the new cell.

In IS-95 inter-frequency handovers and older analog systems such as NMT it will typically be impossible to test the target channel directly while communicating. In this case, other techniques have to be used such as pilot beacons in IS-95. This means that there is almost always a brief break in the communication while searching for the new channel followed by the risk of an unexpected return to the old channel.

If there is no ongoing communication or the communication can be interrupted, it is possible for the mobile unit to spontaneously move from one cell to another and then notify the base station with the strongest signal.

Cellular Frequency Choice in Mobile Phone Networks

The effect of frequency on cell coverage means that different frequencies serve better for different uses. Low frequencies, such as 450 MHz NMT, serve very well for countryside coverage. GSM 900 (900 MHz) is a suitable solution for light urban coverage. GSM 1800 (1.8 GHz) starts to be limited by structural walls. UMTS, at 2.1 GHz is quite similar in coverage to GSM 1800.

Higher frequencies are a disadvantage when it comes to coverage, but it is a decided advantage when it comes to capacity. Picocells, covering e.g. one floor of a building, become possible, and the same frequency can be used for cells which are practically neighbors.

Cell service area may also vary due to interference from transmitting systems, both within and around that cell. This is true especially in CDMA based systems. The receiver requires a certain signal-to-noise ratio, and the transmitter should not send with too high transmission power in view to not cause interference with other transmitters. As the receiver moves away from the transmitter, the power received decreases, so the power control algorithm of the transmitter increases the power it transmits to restore the level of received power. As the interference (noise) rises above the received power from the transmitter, and the power of the transmitter cannot be increased anymore, the signal becomes corrupted and eventually unusable. In CDMA-based systems, the effect of interference from other mobile transmitters in the same cell on coverage area is very marked and has a special name, cell breathing.

One can see examples of cell coverage by studying some of the coverage maps provided by real operators on their web sites or by looking at independently crowdsourced maps such as OpenSignal. In certain cases they may mark the site of the transmitter, in others, it can be calculated by working out the point of strongest coverage.

A cellular repeater is used to extend cell coverage into larger areas. They range from wideband repeaters for consumer use in homes and offices to smart or digital repeaters for industrial needs.

Coverage Comparison of Different Frequencies

The following table shows the dependency of the coverage area of one cell on the frequency of a CDMA2000 network:

Frequency (MHz)	Cell radius (km)	Cell area (km^2)	Relative Cell Count
450	48.9	7521	1
950	26.9	2269	3.3
1800	14.0	618	12.2
2100	12.0	449	16.2

COMPUTER NETWORK

A computer network is a digital telecommunications network which allows nodes to share resources. In computer networks, computing devices exchange data with each other using connections (data links) between nodes. These data links are established over cable media such as wires or optic cables, or wireless media such as Wi-Fi.

Network computer devices that originate, route and terminate the data are called network nodes. Nodes are generally identified by network addresses, and can include hosts such as personal computers, phones, and servers, as well as networking hardware such as routers and switches. Two such devices can be said to be networked together when one device is able to exchange information with the other device, whether or not they have a direct connection to each other. In most cases, application-specific communications protocols are layered (i.e. carried as payload) over other more general communications protocols. This formidable collection of information technology requires skilled network management to keep it all running reliably.

Computer networks support an enormous number of applications and services such as access to the World Wide Web, digital video, digital audio, shared use of application and storage servers, printers, and fax machines, and use of email and instant messaging applications as well as many others. Computer networks differ in the transmission medium used to carry their signals, communications protocols to organize network traffic, the network's size, topology, traffic control mechanism and organizational intent. The best-known computer network is the Internet.

Properties

Computer networking may be considered a branch of electrical engineering, electronics engineering, telecommunications, computer science, information technology or computer engineering, since it relies upon the theoretical and practical application of the related disciplines.

A computer network facilitates interpersonal communications allowing users to communicate efficiently and easily via various means: email, instant messaging, online chat, telephone, video telephone calls, and video conferencing. A network allows sharing of network and computing resources. Users may access and use resources provided by devices on the network, such as printing a document on a shared network printer or use of a shared storage device. A network allows sharing of files, data, and other types of information giving authorized users the ability to access information stored on other computers on the network. Distributed computing uses computing resources across a network to accomplish tasks.

A computer network may be used by security hackers to deploy computer viruses or computer worms on devices connected to the network, or to prevent these devices from accessing the network via a denial-of-service attack.

Network Packet

Computer communication links that do not support packets, such as traditional point-to-point telecommunication links, simply transmit data as a bit stream. However, the overwhelming majority of computer networks carry their data in packets. A network packet is a formatted unit of data (a list of bits or bytes, usually a few tens of bytes to a few kilobytes long) carried by a packet-switched network. Packets are sent through the network to their destination. Once the packets arrive they are reassembled into their original message.

Packets consist of two kinds of data: control information, and user data (payload). The control information provides data the network needs to deliver the user data, for example: source and destination network addresses, error detection codes, and sequencing information. Typically, control information is found in packet headers and trailers, with payload data in between.

With packets, the bandwidth of the transmission medium can be better shared among users than if the network were circuit switched. When one user is not sending packets, the link can be filled with packets from other users, and so the cost can be shared, with relatively little interference, provided the link isn't overused. Often the route a packet needs to take through a network is not immediately available. In that case the packet is queued and waits until a link is free.

Network Topology

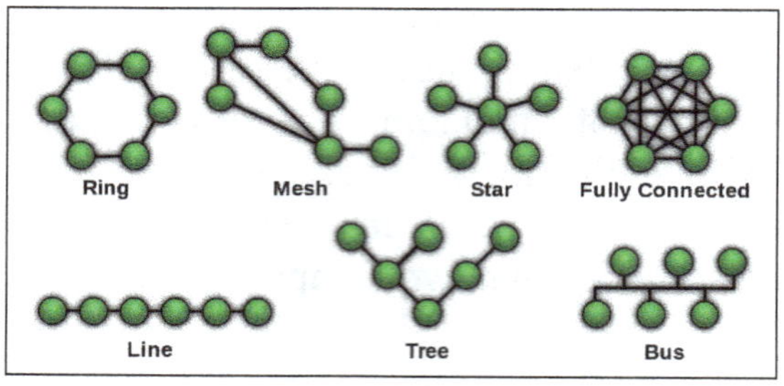

Common network topologies.

The physical layout of a network is usually less important than the topology that connects network nodes. Most diagrams that describe a physical network are therefore topological, rather than geographic. The symbols on these diagrams usually denote network links and network nodes.

Network topology is the layout or organizational hierarchy of interconnected nodes of a computer network. Different network topologies can affect throughput, but reliability is often more critical. With many technologies, such as bus networks, a single failure can cause the network to fail entirely. In general the more interconnections there are, the more robust the network is; but the more expensive it is to install.

Common layouts are:

- A bus network: All nodes are connected to a common medium along this medium. This was the layout used in the original Ethernet, called 10BASE5 and 10BASE2. This is still a common topology on the data link layer, although modern physical layer variants use point-to-point links instead.
- A star network: All nodes are connected to a special central node. This is the typical layout found in a Wireless LAN, where each wireless client connects to the central Wireless access point.
- A ring network: Each node is connected to its left and right neighbour node, such that all nodes are connected and that each node can reach each other node by traversing nodes left- or rightwards. The Fiber Distributed Data Interface (FDDI) made use of such a topology.
- A mesh network: Each node is connected to an arbitrary number of neighbours in such a way that there is at least one traversal from any node to any other.
- A fully connected network: Each node is connected to every other node in the network.
- A tree network: Nodes are arranged hierarchically.

The physical layout of the nodes in a network may not necessarily reflect the network topology. As an example, with FDDI, the network topology is a ring (actually two counter-rotating rings), but the physical topology is often a star, because all neighboring connections can be routed via a central physical location. Physical layout is not completely irrelevant, however, as common ducting and equipment locations can represent single points of failure due to issues like fires, power failures and flooding.

Overlay Network

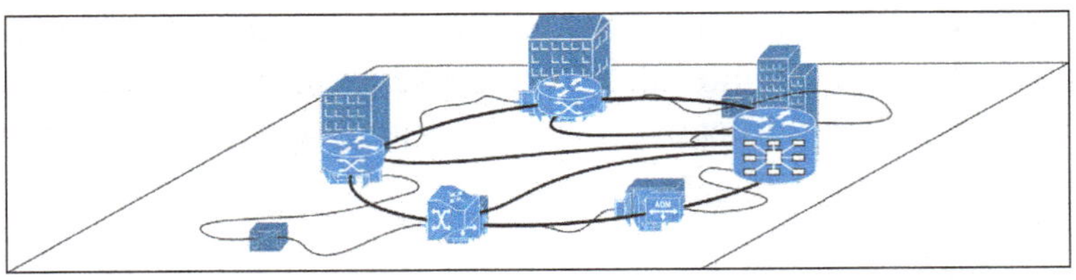

A sample overlay network.

An overlay network is a virtual computer network that is built on top of another network. Nodes in the overlay network are connected by virtual or logical links. Each link corresponds to a path, perhaps through many physical links, in the underlying network. The topology of the overlay network may (and often does) differ from that of the underlying one. For example, many peer-to-peer networks are overlay networks. They are organized as nodes of a virtual system of links that run on top of the Internet.

Overlay networks have been around since the invention of networking when computer systems were connected over telephone lines using modems, before any data network existed.

The most striking example of an overlay network is the Internet itself. The Internet itself was initially built as an overlay on the telephone network. Even today, each Internet node can communicate with virtually any other through an underlying mesh of sub-networks of wildly different topologies and technologies. Address resolution and routing are the means that allow mapping of a fully connected IP overlay network to its underlying network.

Another example of an overlay network is a distributed hash table, which maps keys to nodes in the network. In this case, the underlying network is an IP network, and the overlay network is a table (actually a map) indexed by keys.

Overlay networks have also been proposed as a way to improve Internet routing, such as through quality of service guarantees to achieve higher-quality streaming media. Previous proposals such as IntServ, DiffServ, and IP Multicast have not seen wide acceptance largely because they require modification of all routers in the network. On the other hand, an overlay network can be incrementally deployed on end-hosts running the overlay protocol software, without cooperation from Internet service providers. The overlay network has no control over how packets are routed in the underlying network between two overlay nodes, but it can control, for example, the sequence of overlay nodes that a message traverses before it reaches its destination.

For example, Akamai Technologies manages an overlay network that provides reliable, efficient content delivery (a kind of multicast). Academic research includes end system multicast, resilient routing and quality of service studies, among others.

Network Links

The transmission media (often referred to in the literature as the *physical medium*) used to link devices to form a computer network include electrical cable, optical fiber, and radio waves. In the OSI model, these are defined at layers 1 and 2 — the physical layer and the data link layer.

A widely adopted family of transmission media used in local area network (LAN) technology is collectively known as Ethernet. The media and protocol standards that enable communication between networked devices over Ethernet are defined by IEEE 802.3. Ethernet transmits data over both copper and fiber cables. Wireless LAN standards use radio waves, others use infrared signals as a transmission medium. Power line communication uses a building's power cabling to transmit data.

Wired Technologies

Fiber optic cables are used to transmit light from one computer/network node to another.

The following classes of wired technologies are used in computer networking.

- Coaxial cable is widely used for cable television systems, office buildings, and other work-sites for local area networks. Transmission speed ranges from 200 million bits per second to more than 500 million bits per second.
- ITU-T G.hn technology uses existing home wiring (coaxial cable, phone lines and power lines) to create a high-speed local area network.
- Twisted pair cabling is used for wired Ethernet and other standards. It typically consists of 4 pairs of copper cabling that can be utilized for both voice and data transmission. The use of two wires twisted together helps to reduce crosstalk and electromagnetic induction. The transmission speed ranges from 2 Mbit/s to 10 Gbit/s. Twisted pair cabling comes in two forms: unshielded twisted pair (UTP) and shielded twisted-pair (STP). Each form comes in several category ratings, designed for use in various scenarios.

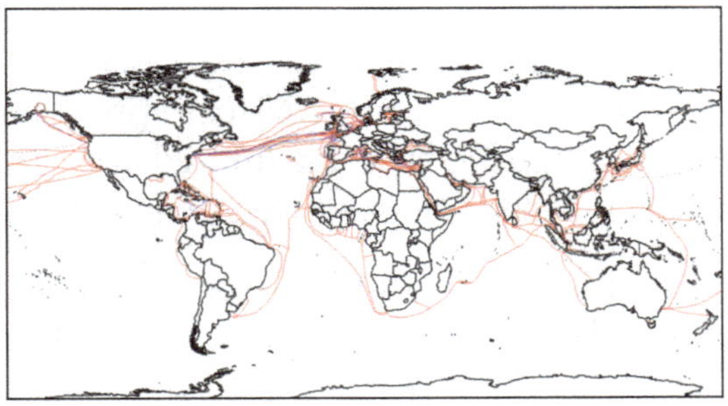

2007 map showing submarine optical fiber telecommunication cables around the world.

- An optical fiber is a glass fiber. It carries pulses of light that represent data. Some advantages of optical fibers over metal wires are very low transmission loss and immunity to electrical interference. Optical fibers can simultaneously carry multiple streams of data on different wavelengths of light, which greatly increases the rate that data can be sent to up to trillions of bits per second. Optic fibers can be used for long runs of cable carrying very high data rates, and are used for undersea cables to interconnect continents. There are two basic types of fiber optics, single-mode optical fiber (SMF) and multi-mode optical fiber (MMF). Single-mode fiber has the advantage of being able to sustain a coherent signal for dozens or even a hundred kilometers. Multimode fiber is cheaper to terminate but is limited to a few hundred or even only a few dozens of meters, depending on the data rate and cable grade.

Wireless Technologies

Computers are very often connected to networks using wireless links.

Network connections can be established wirelessly using radio or other electromagnetic means of communication.

- Terrestrial microwave: Terrestrial microwave communication uses Earth-based transmitters and receivers resembling satellite dishes. Terrestrial microwaves are in the low gigahertz range, which limits all communications to line-of-sight. Relay stations are spaced approximately 40 miles (64 km) apart.
- Communications satellites: Satellites communicate also communicate via microwave. The satellites are stationed in space, typically in geosynchronous orbit 35,400 km (22,000 mi) above the equator. These Earth-orbiting systems are capable of receiving and relaying voice, data, and TV signals.
- Cellular networks use several radio communications technologies. The systems divide the region covered into multiple geographic areas. Each area is served by a low-power transceiver.
- Radio and spread spectrum technologies: Wireless LANs use a high-frequency radio technology similar to digital cellular. Wireless LANs use spread spectrum technology to enable communication between multiple devices in a limited area. IEEE 802.11 defines a common flavor of open-standards wireless radio-wave technology known as Wi-Fi.
- Free-space optical communication uses visible or invisible light for communications. In most cases, line-of-sight propagation is used, which limits the physical positioning of communicating devices.

Exotic Technologies

There have been various attempts at transporting data over exotic media:

- IP over Avian Carriers was a humorous April fool's Request for Comments, issued as RFC 1149. It was implemented in real life in 2001.
- Extending the Internet to interplanetary dimensions via radio waves, the Interplanetary Internet.

Both cases have a large round-trip delay time, which gives slow two-way communication, but doesn't prevent sending large amounts of information.

Network Nodes

Apart from any physical transmission media there may be, networks comprise additional basic system building blocks, such as network interface controllers (NICs), repeaters, hubs, bridges, switches, routers, modems, and firewalls. Any particular piece of equipment will frequently contain multiple building blocks and perform multiple functions.

Network Interfaces

A network interface controller (NIC) is computer hardware that provides a computer with the ability to access the transmission media, and has the ability to process low-level network information.

For example, the NIC may have a connector for accepting a cable, or an aerial for wireless transmission and reception, and the associated circuitry.

An ATM network interface in the form of an accessory card. A lot of network interfaces are built-in.

The NIC responds to traffic addressed to a network address for either the NIC or the computer as a whole.

In Ethernet networks, each network interface controller has a unique Media Access Control (MAC) address—usually stored in the controller's permanent memory. To avoid address conflicts between network devices, the Institute of Electrical and Electronics Engineers (IEEE) maintains and administers MAC address uniqueness. The size of an Ethernet MAC address is six octets. The three most significant octets are reserved to identify NIC manufacturers. These manufacturers, using only their assigned prefixes, uniquely assign the three least-significant octets of every Ethernet interface they produce.

Repeaters and Hubs

A repeater is an electronic device that receives a network signal, cleans it of unnecessary noise and regenerates it. The signal is retransmitted at a higher power level, or to the other side of an obstruction, so that the signal can cover longer distances without degradation. In most twisted pair Ethernet configurations, repeaters are required for cable that runs longer than 100 meters. With fiber optics, repeaters can be tens or even hundreds of kilometers apart.

A repeater with multiple ports is known as an Ethernet hub. Repeaters work on the physical layer of the OSI model. Repeaters require a small amount of time to regenerate the signal. This can cause a propagation delay that affects network performance and may affect proper function. As a result, many network architectures limit the number of repeaters that can be used in a row, e.g., the Ethernet 5-4-3 rule. Hubs and repeaters in LANs have been mostly obsoleted by modern switches.

Bridges

A network bridge connects and filters traffic between two network segments at the data link layer (layer 2) of the OSI model to form a single network. This breaks the network's collision domain but maintains a unified broadcast domain. Network segmentation breaks down a large, congested network into an aggregation of smaller, more efficient networks.

Bridges come in three basic types:

- Local bridges: Directly connect LANs
- Remote bridges: Can be used to create a wide area network (WAN) link between LANs. Remote bridges, where the connecting link is slower than the end networks, largely have been replaced with routers.
- Wireless bridges: Can be used to join LANs or connect remote devices to LANs.

Switches

A network switch is a device that forwards and filters OSI layer 2 datagrams (frames) between ports based on the destination MAC address in each frame. A switch is distinct from a hub in that it only forwards the frames to the physical ports involved in the communication rather than all ports connected. It can be thought of as a multi-port bridge. It learns to associate physical ports to MAC addresses by examining the source addresses of received frames. If an unknown destination is targeted, the switch broadcasts to all ports but the source. Switches normally have numerous ports, facilitating a star topology for devices, and cascading additional switches.

Routers

A typical home or small office router showing the ADSL telephone line and Ethernet network cable connections.

A router is an internetworking device that forwards packets between networks by processing the routing information included in the packet or datagram (Internet protocol information from layer 3). The routing information is often processed in conjunction with the routing table (or forwarding table). A router uses its routing table to determine where to forward packets. A destination in a routing table can include a "null" interface, also known as the "black hole" interface because data can go into it, however, no further processing is done for said data, i.e. the packets are dropped.

Modems

Modems (Modulator-DEModulator) are used to connect network nodes via wire not originally designed for digital network traffic, or for wireless. To do this one or more carrier signals are modulated by the digital signal to produce an analog signal that can be tailored to give the required properties for transmission. Modems are commonly used for telephone lines, using a digital subscriber line technology.

Firewalls

A firewall is a network device for controlling network security and access rules. Firewalls are typically configured to reject access requests from unrecognized sources while allowing actions from recognized ones. The vital role firewalls play in network security grows in parallel with the constant increase in cyber attacks.

Communication Protocols

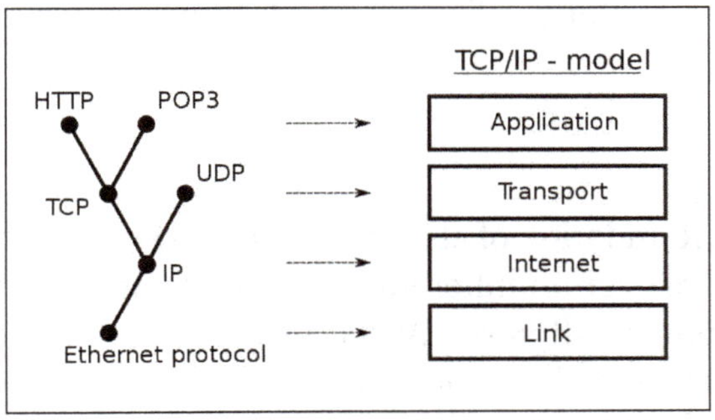

The TCP/IP model or Internet layering scheme and its relation to common protocols often layered on top of it.

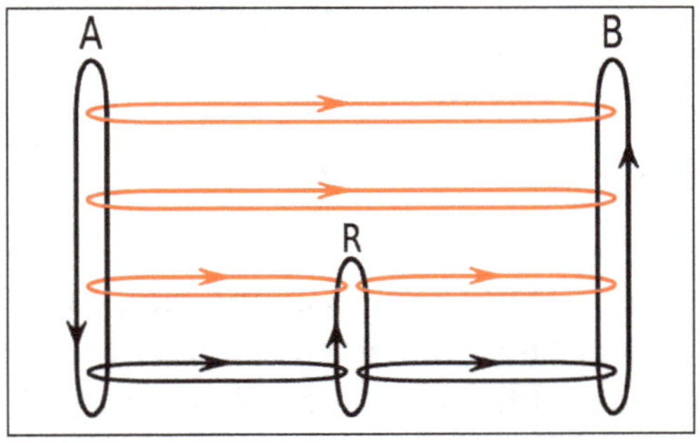

Message flows (A-B) in the presence of a router (R), red flows are effective communication paths, black paths are across the actual network links.

A communication protocol is a set of rules for exchanging information over a network. In a protocol stack, each protocol leverages the services of the protocol layer below it, until the lowest layer controls the hardware which sends information across the media. The use of protocol layering is today ubiquitous across the field of computer networking. An important example of a protocol stack is HTTP (the World Wide Web protocol) running over TCP over IP (the Internet protocols) over IEEE 802.11 (the Wi-Fi protocol). This stack is used between the wireless router and the home user's personal computer when the user is surfing the web.

Communication protocols have various characteristics. They may be connection-oriented or connectionless, they may use circuit mode or packet switching, and they may use hierarchical addressing or flat addressing.

There are many communication protocols, a few of which are described below:

IEEE 802

IEEE 802 is a family of IEEE standards dealing with local area networks and metropolitan area networks. The complete IEEE 802 protocol suite provides a diverse set of networking capabilities. The protocols have a flat addressing scheme. They operate mostly at levels 1 and 2 of the OSI model.

For example, MAC bridging (IEEE 802.1D) deals with the routing of Ethernet packets using a Spanning Tree Protocol. IEEE 802.1Q describes VLANs, and IEEE 802.1X defines a port-based Network Access Control protocol, which forms the basis for the authentication mechanisms used in VLANs (but it is also found in WLANs) – it is what the home user sees when the user has to enter a "wireless access key".

Ethernet

Ethernet, sometimes simply called *LAN*, is a family of protocols used in wired LANs, described by a set of standards together called IEEE 802.3 published by the Institute of Electrical and Electronics Engineers.

Wireless LAN

Wireless LAN, also widely known as WLAN or WiFi, is probably the most well-known member of the IEEE 802 protocol family for home users today. It is standardized by IEEE 802.11 and shares many properties with wired Ethernet.

Internet Protocol Suite

The Internet Protocol Suite, also called TCP/IP, is the foundation of all modern networking. It offers connection-less as well as connection-oriented services over an inherently unreliable network traversed by data-gram transmission at the Internet protocol (IP) level. At its core, the protocol suite defines the addressing, identification, and routing specifications for Internet Protocol Version 4 (IPv4) and for IPv6, the next generation of the protocol with a much enlarged addressing capability.

SONET/SDH

Synchronous optical networking (SONET) and Synchronous Digital Hierarchy (SDH) are standardized multiplexing protocols that transfer multiple digital bit streams over optical fiber using lasers. They were originally designed to transport circuit mode communications from a variety of different sources, primarily to support real-time, uncompressed, circuit-switched voice encoded in PCM (Pulse-Code Modulation) format. However, due to its protocol neutrality and transport-oriented features, SONET/SDH also was the obvious choice for transporting Asynchronous Transfer Mode (ATM) frames.

Asynchronous Transfer Mode

Asynchronous Transfer Mode (ATM) is a switching technique for telecommunication networks. It uses asynchronous time-division multiplexing and encodes data into small, fixed-sized cells. This differs from other protocols such as the Internet Protocol Suite or Ethernet that use variable sized packets or frames. ATM has similarity with both circuit and packet switched networking. This makes it a good choice for a network that must handle both traditional high-throughput data traffic, and real-time, low-latency content such as voice and video. ATM uses a connection-oriented model in which a virtual circuit must be established between two endpoints before the actual data exchange begins.

While the role of ATM is diminishing in favor of next-generation networks, it still plays a role in the last mile, which is the connection between an Internet service provider and the home user.

Cellular Standards

There are a number of different digital cellular standards, including: Global System for Mobile Communications (GSM), General Packet Radio Service (GPRS), cdmaOne, CDMA2000,

Evolution-Data Optimized (EV-DO), Enhanced Data Rates for GSM Evolution (EDGE), Universal Mobile Telecommunications System (UMTS), Digital Enhanced Cordless Telecommunications (DECT), Digital AMPS (IS-136/TDMA), and Integrated Digital Enhanced Network (iDEN).

Geographic Scale

A network can be characterized by its physical capacity or its organizational purpose. Use of the network, including user authorization and access rights, differ accordingly.

Nanoscale Network

A nanoscale communication network has key components implemented at the nanoscale including message carriers and leverages physical principles that differ from macroscale communication mechanisms. Nanoscale communication extends communication to very small sensors and actuators such as those found in biological systems and also tends to operate in environments that would be too harsh for classical communication.

Personal Area Network

A personal area network (PAN) is a computer network used for communication among computer and different information technological devices close to one person. Some examples of devices that are used in a PAN are personal computers, printers, fax machines, telephones, PDAs, scanners, and even video game consoles. A PAN may include wired and wireless devices. The reach of a PAN typically extends to 10 meters. A wired PAN is usually constructed with USB and FireWire connections while technologies such as Bluetooth and infrared communication typically form a wireless PAN.

Local Area Network

A local area network (LAN) is a network that connects computers and devices in a limited geographical area such as a home, school, office building, or closely positioned group of buildings. Each computer or device on the network is a node. Wired LANs are most likely based on Ethernet technology. Newer standards such as ITU-T G.hn also provide a way to create a wired LAN using existing wiring, such as coaxial cables, telephone lines, and power lines.

The defining characteristics of a LAN, in contrast to a wide area network (WAN), include higher data transfer rates, limited geographic range, and lack of reliance on leased lines to provide connectivity. Current Ethernet or other IEEE 802.3 LAN technologies operate at data transfer rates up to 100 Gbit/s, standardized by IEEE in 2010. Currently, 400 Gbit/s Ethernet is being developed.

A LAN can be connected to a WAN using a router.

Home Area Network

A home area network (HAN) is a residential LAN used for communication between digital devices typically deployed in the home, usually a small number of personal computers and accessories, such as printers and mobile computing devices. An important function is the sharing of Internet access, often a broadband service through a cable TV or digital subscriber line (DSL) provider.

Storage Area Network

A storage area network (SAN) is a dedicated network that provides access to consolidated, block level data storage. SANs are primarily used to make storage devices, such as disk arrays, tape libraries, and optical jukeboxes, accessible to servers so that the devices appear like locally attached devices to the operating system. A SAN typically has its own network of storage devices that are generally not accessible through the local area network by other devices. The cost and complexity of SANs dropped in the early 2000s to levels allowing wider adoption across both enterprise and small to medium-sized business environments.

Campus Area Network

A campus area network (CAN) is made up of an interconnection of LANs within a limited geographical area. The networking equipment (switches, routers) and transmission media (optical fiber, copper plant, Cat5 cabling, etc.) are almost entirely owned by the campus tenant/owner (an enterprise, university, government, etc).

For example, a university campus network is likely to link a variety of campus buildings to connect academic colleges or departments, the library, and student residence halls.

Backbone Network

A backbone network is part of a computer network infrastructure that provides a path for the exchange of information between different LANs or sub-networks. A backbone can tie together diverse networks within the same building, across different buildings, or over a wide area.

For example, a large company might implement a backbone network to connect departments that are located around the world. The equipment that ties together the departmental networks constitutes the network backbone. When designing a network backbone, network performance and network congestion are critical factors to take into account. Normally, the backbone network's capacity is greater than that of the individual networks connected to it.

Another example of a backbone network is the Internet backbone, which is the set of wide area networks (WANs) and core routers that tie together all networks connected to the Internet.

Metropolitan Area Network

A Metropolitan area network (MAN) is a large computer network that usually spans a city or a large campus.

Wide Area Network

A wide area network (WAN) is a computer network that covers a large geographic area such as a city, country, or spans even intercontinental distances. A WAN uses a communications channel that combines many types of media such as telephone lines, cables, and air waves. A WAN often makes use of transmission facilities provided by common carriers, such as telephone companies. WAN technologies generally function at the lower three layers of the OSI reference model: the physical layer, the data link layer, and the network layer.

Enterprise Private Network

An enterprise private network is a network that a single organization builds to interconnect its office locations (e.g., production sites, head offices, remote offices, shops) so they can share computer resources.

Virtual Private Network

A virtual private network (VPN) is an overlay network in which some of the links between nodes are carried by open connections or virtual circuits in some larger network (e.g., the Internet) instead of by physical wires. The data link layer protocols of the virtual network are said to be tunneled through the larger network when this is the case. One common application is secure communications through the public Internet, but a VPN need not have explicit security features, such as authentication or content encryption. VPNs, for example, can be used to separate the traffic of different user communities over an underlying network with strong security features.

VPN may have best-effort performance, or may have a defined service level agreement (SLA) between the VPN customer and the VPN service provider. Generally, a VPN has a topology more complex than point-to-point.

Global Area Network

A global area network (GAN) is a network used for supporting mobile across an arbitrary number of wireless LANs, satellite coverage areas, etc. The key challenge in mobile communications is handing off user communications from one local coverage area to the next. In IEEE Project 802, this involves a succession of terrestrial wireless LANs.

Organizational Scope

Networks are typically managed by the organizations that own them. Private enterprise networks may use a combination of intranets and extranets. They may also provide network access to the Internet, which has no single owner and permits virtually unlimited global connectivity.

Intranet

An intranet is a set of networks that are under the control of a single administrative entity. The intranet uses the IP protocol and IP-based tools such as web browsers and file transfer applications. The administrative entity limits use of the intranet to its authorized users. Most commonly, an intranet is the internal LAN of an organization. A large intranet typically has at least one web server to provide users with organizational information. An intranet is also anything behind the router on a local area network.

Extranet

An extranet is a network that is also under the administrative control of a single organization, but supports a limited connection to a specific external network. For example, an organization may provide access to some aspects of its intranet to share data with its business partners or customers.

These other entities are not necessarily trusted from a security standpoint. Network connection to an extranet is often, but not always, implemented via WAN technology.

Internetwork

An internetwork is the connection of multiple computer networks via a common routing technology using routers.

Internet

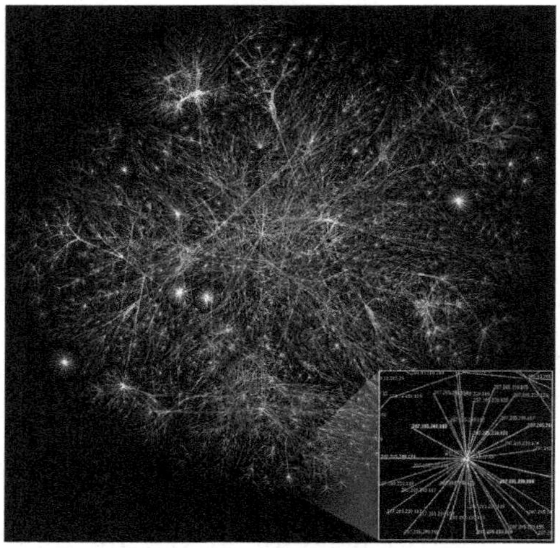

Partial map of the Internet based on the January 15, 2005 data found on opte.org. Each line is drawn between two nodes, representing two IP addresses. The length of the lines are indicative of the delay between those two nodes. This graph represents less than 30% of the Class C networks reachable.

The Internet is the largest example of an internetwork. It is a global system of interconnected governmental, academic, corporate, public, and private computer networks. It is based on the networking technologies of the Internet Protocol Suite. It is the successor of the Advanced Research Projects Agency Network (ARPANET) developed by DARPA of the United States Department of Defense. The Internet is also the communications backbone underlying the World Wide Web (WWW).

Participants in the Internet use a diverse array of methods of several hundred documented, and often standardized, protocols compatible with the Internet Protocol Suite and an addressing system (IP addresses) administered by the Internet Assigned Numbers Authority and address registries. Service providers and large enterprises exchange information about the reachability of their address spaces through the Border Gateway Protocol (BGP), forming a redundant worldwide mesh of transmission paths.

Darknet

A darknet is an overlay network, typically running on the Internet, that is only accessible through specialized software. A darknet is an anonymizing network where connections are made only between trusted peers — sometimes called "friends" (F2F) — using non-standard protocols and ports.

Darknets are distinct from other distributed peer-to-peer networks as sharing is anonymous (that is, IP addresses are not publicly shared), and therefore users can communicate with little fear of governmental or corporate interference.

Routing

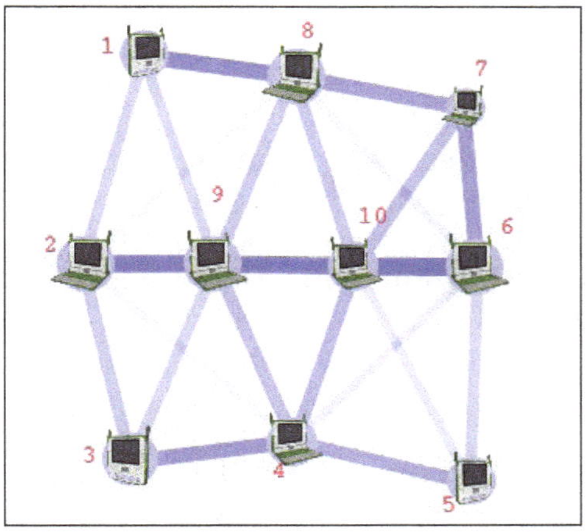

Routing calculates good paths through a network for information to take. For example, from node 1 to node 6 the best routes are likely to be 1-8-7-6 or 1-8-10-6, as this has the thickest routes.

Routing is the process of selecting network paths to carry network traffic. Routing is performed for many kinds of networks, including circuit switching networks and packet switched networks.

In packet switched networks, routing directs packet forwarding (the transit of logically addressed network packets from their source toward their ultimate destination) through intermediate nodes. Intermediate nodes are typically network hardware devices such as routers, bridges, gateways, firewalls, or switches. General-purpose computers can also forward packets and perform routing, though they are not specialized hardware and may suffer from limited performance. The routing process usually directs forwarding on the basis of routing tables, which maintain a record of the routes to various network destinations. Thus, constructing routing tables, which are held in the router's memory, is very important for efficient routing.

There are usually multiple routes that can be taken, and to choose between them, different elements can be considered to decide which routes get installed into the routing table, such as (sorted by priority):

- Prefix-Length: Where longer subnet masks are preferred (independent if it is within a routing protocol or over different routing protocol).
- Metric: Where a lower metric/cost is preferred (only valid within one and the same routing protocol).
- Administrative distance: Where a lower distance is preferred (only valid between different routing protocols).

Most routing algorithms use only one network path at a time. Multipath routing techniques enable the use of multiple alternative paths.

Routing, in a more narrow sense of the term, is often contrasted with bridging in its assumption that network addresses are structured and that similar addresses imply proximity within the network. Structured addresses allow a single routing table entry to represent the route to a group of devices. In large networks, structured addressing (routing, in the narrow sense) outperforms unstructured addressing (bridging). Routing has become the dominant form of addressing on the Internet. Bridging is still widely used within localized environments.

Network Service

Network services are applications hosted by servers on a computer network, to provide some functionality for members or users of the network, or to help the network itself to operate.

The World Wide Web, E-mail, printing and network file sharing are examples of well-known network services. Network services such as DNS (Domain Name System) give names for IP and MAC addresses (people remember names like "nm.lan" better than numbers like "210.121.67.18"), and DHCP to ensure that the equipment on the network has a valid IP address.

Services are usually based on a service protocol that defines the format and sequencing of messages between clients and servers of that network service.

Network Performance

Quality of Service

Depending on the installation requirements, network performance is usually measured by the quality of service of a telecommunications product. The parameters that affect this typically can include throughput, jitter, bit error rate and latency.

The following list gives examples of network performance measures for a circuit-switched network and one type of packet-switched network, viz. ATM:

- Circuit-switched networks: In circuit switched networks, network performance is synonymous with the grade of service. The number of rejected calls is a measure of how well the network is performing under heavy traffic loads. Other types of performance measures can include the level of noise and echo.
- ATM: In an Asynchronous Transfer Mode (ATM) network, performance can be measured by line rate, quality of service (QoS), data throughput, connect time, stability, technology, modulation technique and modem enhancements.

There are many ways to measure the performance of a network, as each network is different in nature and design. Performance can also be modelled instead of measured. For example, state transition diagrams are often used to model queuing performance in a circuit-switched network. The network planner uses these diagrams to analyze how the network performs in each state, ensuring that the network is optimally designed.

Network Congestion

Network congestion occurs when a link or node is subjected to a greater data load than it is rated for, resulting in a deterioration of its quality of service. Typical effects include queueing delay, packet loss or the blocking of new connections. A consequence of these latter two is that incremental increases in offered load lead either to only a small increase in network throughput, or to a reduction in network throughput.

Network protocols that use aggressive retransmissions to compensate for packet loss tend to keep systems in a state of network congestion—even after the initial load is reduced to a level that would not normally induce network congestion. Thus, networks using these protocols can exhibit two stable states under the same level of load. The stable state with low throughput is known as congestive collapse.

Modern networks use congestion control, congestion avoidance and traffic control techniques to try to avoid congestion collapse. These include: exponential backoff in protocols such as 802.11's CSMA/CA and the original Ethernet, window reduction in TCP, and fair queueing in devices such as routers. Another method to avoid the negative effects of network congestion is implementing priority schemes, so that some packets are transmitted with higher priority than others. Priority schemes do not solve network congestion by themselves, but they help to alleviate the effects of congestion for some services. An example of this is 802.1p. A third method to avoid network congestion is the explicit allocation of network resources to specific flows. One example of this is the use of Contention-Free Transmission Opportunities (CFTXOPs) in the ITU-T G.hn standard, which provides high-speed (up to 1 Gbit/s) Local area networking over existing home wires (power lines, phone lines and coaxial cables).

For the Internet, RFC 2914 addresses the subject of congestion control in detail.

Network Resilience

Network resilience is "the ability to provide and maintain an acceptable level of service in the face of faults and challenges to normal operation."

Security

Network Security

Network security consists of provisions and policies adopted by the network administrator to prevent and monitor unauthorized access, misuse, modification, or denial of the computer network and its network-accessible resources. Network security is the authorization of access to data in a network, which is controlled by the network administrator. Users are assigned an ID and password that allows them access to information and programs within their authority. Network security is used on a variety of computer networks, both public and private, to secure daily transactions and communications among businesses, government agencies and individuals.

Network Surveillance

Network surveillance is the monitoring of data being transferred over computer networks such as the Internet. The monitoring is often done surreptitiously and may be done by or at the behest of governments, by corporations, criminal organizations, or individuals. It may or may not be legal and may or may not require authorization from a court or other independent agency.

Computer and network surveillance programs are widespread today, and almost all Internet traffic is or could potentially be monitored for clues to illegal activity.

Surveillance is very useful to governments and law enforcement to maintain social control, recognize and monitor threats, and prevent/investigate criminal activity. With the advent of programs such as the Total Information Awareness program, technologies such as high speed surveillance computers and biometrics software, and laws such as the Communications Assistance For Law Enforcement Act, governments now possess an unprecedented ability to monitor the activities of citizens.

However, many civil rights and privacy groups—such as Reporters Without Borders, the Electronic Frontier Foundation, and the American Civil Liberties Union—have expressed concern that increasing surveillance of citizens may lead to a mass surveillance society, with limited political and personal freedoms. Fears such as this have led to numerous lawsuits such as *Hepting v. AT&T*. The hacktivist group Anonymous has hacked into government websites in protest of what it considers "draconian surveillance".

End-to-end Encryption

End-to-end encryption (E2EE) is a digital communications paradigm of uninterrupted protection of data traveling between two communicating parties. It involves the originating party encrypting data so only the intended recipient can decrypt it, with no dependency on third parties. End-to-end encryption prevents intermediaries, such as Internet providers or application service providers, from discovering or tampering with communications. End-to-end encryption generally protects both confidentiality and integrity.

Examples of end-to-end encryption include HTTPS for web traffic, PGP for email, OTR for instant messaging, ZRTP for telephony, and TETRA for radio.

Typical server-based communications systems do not include end-to-end encryption. These systems can only guarantee protection of communications between clients and servers, not between the communicating parties themselves. Examples of non-E2EE systems are Google Talk, Yahoo Messenger, Facebook, and Dropbox. Some such systems, for example LavaBit and SecretInk, have even described themselves as offering "end-to-end" encryption when they do not. Some systems that normally offer end-to-end encryption have turned out to contain a back door that subverts negotiation of the encryption key between the communicating parties, for example Skype or Hushmail.

The end-to-end encryption paradigm does not directly address risks at the communications endpoints themselves, such as the technical exploitation of clients, poor quality random number generators, or key escrow. E2EE also does not address traffic analysis, which relates to things such as the identities of the end points and the times and quantities of messages that are sent.

SSL/TLS

The introduction and rapid growth of e-commerce on the World Wide Web in the mid-1990s made it obvious that some form of authentication and encryption was needed. Netscape took the first shot at a new standard. At the time, the dominant web browser was Netscape Navigator. Netscape created a standard called secure socket layer (SSL). SSL requires a server with a certificate. When a client requests access to an SSL-secured server, the server sends a copy of the certificate to the client. The SSL client checks this certificate (all web browsers come with an exhaustive list of CA root certificates preloaded), and if the certificate checks out, the server is authenticated and the client negotiates a symmetric-key cipher for use in the session. The session is now in a very secure encrypted tunnel between the SSL server and the SSL client.

Views of Networks

Users and network administrators typically have different views of their networks. Users can share printers and some servers from a workgroup, which usually means they are in the same geographic location and are on the same LAN, whereas a Network Administrator is responsible to keep that network up and running. A community of interest has less of a connection of being in a local area, and should be thought of as a set of arbitrarily located users who share a set of servers, and possibly also communicate via peer-to-peer technologies.

Network administrators can see networks from both physical and logical perspectives. The physical perspective involves geographic locations, physical cabling, and the network elements (e.g., routers, bridges and application layer gateways) that interconnect via the transmission media. Logical networks, called, in the TCP/IP architecture, subnets, map onto one or more transmission media. For example, a common practice in a campus of buildings is to make a set of LAN cables in each building appear to be a common subnet, using virtual LAN (VLAN) technology.

Both users and administrators are aware, to varying extents, of the trust and scope characteristics of a network. Again using TCP/IP architectural terminology, an intranet is a community of interest under private administration usually by an enterprise, and is only accessible by authorized users (e.g. employees). Intranets do not have to be connected to the Internet, but generally have a limited connection. An extranet is an extension of an intranet that allows secure communications to users outside of the intranet (e.g. business partners, customers).

Unofficially, the Internet is the set of users, enterprises, and content providers that are interconnected by Internet Service Providers (ISP). From an engineering viewpoint, the Internet is the set of subnets, and aggregates of subnets, which share the registered IP address space and exchange information about the reachability of those IP addresses using the Border Gateway Protocol. Typically, the human-readable names of servers are translated to IP addresses, transparently to users, via the directory function of the Domain Name System (DNS).

Over the Internet, there can be business-to-business (B2B), business-to-consumer (B2C) and consumer-to-consumer (C2C) communications. When money or sensitive information is exchanged, the communications are apt to be protected by some form of communications security mechanism. Intranets and extranets can be securely superimposed onto the Internet, without any access by general Internet users and administrators, using secure Virtual Private Network (VPN) technology.

ARPANET

ARPANET (Advanced Research Projects Agency Network) was an experimental computer network that was the forerunner of the Internet. The Advanced Research Projects Agency (ARPA), an arm of the U.S. Defense Department, funded the development of the Advanced Research Projects Agency Network (ARPANET) in the late 1960s. Its initial purpose was to link computers at Pentagon-funded research institutions over telephone lines.

At the height of the Cold War, military commanders were seeking a computer communications system without a central core, with no headquarters or base of operations that could be attacked and destroyed by enemies thus blacking out the entire network in one fell swoop. ARPANET's purpose was always more academic than military, but, as more academic facilities connected to it, the network did take on the tentacle-like structure military officials had envisioned. The Internet essentially retains that form, although on a much larger scale.

Roots of a Network

ARPANET was an end-product of a decade of computer-communications developments spurred by military concerns that the Soviets might use their jet bombers to launch surprise nuclear attacks against the United States. By the 1960s, a system called SAGE (Semi-Automatic Ground Environment) had already been built and was using computers to track incoming enemy aircraft and to coordinate military response. The system included 23 "direction centers," each with a massive mainframe computer that could track 400 planes, distinguishing friendly aircraft from enemy bombers. The system required six years and $61 billion to implement.

The system's name hints at its importance, as author John Naughton points out. The system was only "semi-automatic," so human interaction was pivotal. For Joseph Carl Robnett Licklider, who would became the first director of ARPA's Information Processing Techniques Office (IPTO), the SAGE network demonstrated above all else the enormous power of interactive computing—or, as he refered to it in a seminal 1960 essay, of "man-computer symbiosis." In his essay, one of the most important in the history of computing, Licklider posited the then-radical belief that a marriage of the human mind with the computer would eventually result in better decision-making.

In 1962, Licklider joined ARPA. According to Naughton, his brief two-year stint at the organization seeded everything that was to follow. His tenure signaled the demilitarization of ARPA; it was Licklider who changed the name of his office from Command and Control Research to IPTO. "Lick," as he insisted on being called, brought to the project an emphasis on interactive computing and the prevalent utopian conviction that humans teamed with computers could create a better world.

Perhaps in part because of Cold War fears, during Licklider's IPTO tenure, it is estimated that 70 percent of all U.S. computer-science research was funded by ARPA. But many of those involved said that the agency was far from being a restrictive militaristic environment and that it gave them free rein to try out radical ideas. As a result, ARPA was the birthplace not only of computer networks and the Internet but also of computer graphics, parallel processing, computer flight simulation, and other key achievements.

Ivan Sutherland succeeded Licklider as IPTO director in 1964, and two years later Robert Taylor became IPTO director. Taylor would become a key figure in ARPANET's development, partly because of his observational abilities. In the Pentagon's IPTO office, Taylor had access to three teletype terminals, each hooked up to one of three remote ARPA-supported time-sharing mainframe computers—at Systems Development Corp. in Santa Monica, at UC Berkeley's Genie Project, and at MIT's Compatible Time-Sharing System project (later known as Multics).

In his room at the Pentagon, Taylor's access to time-shared systems led him to a key social observation. He could watch as computers at all three remote facilities came alive with activity, connecting local users. Time-shared computers allowed people to exchange messages and share files. Through the computers, people could learn about each other. Interactive communities formed around the machines.

Taylor also decided that it made no sense to require three teletype machines just to communicate with three incompatible computer systems. It would be much more efficient if the three were merged into one, with a single computer-language protocol that could allow any terminal to communicate with any other terminal. These insights led Taylor to propose and secure funding for ARPANET.

A plan for the network was first made available publicly in October 1967, at an Association for Computing Machinery (ACM) symposium in Gatlinburg, Tennessee. There, plans were announced for building a computer network that would link 16 ARPA-sponsored universities and research centers across the United States. In the summer of 1968, the Defense Department put out a call for competitive bids to build the network, and in January 1969 Bolt, Beranek, and Newman (BBN) of Cambridge, Massachusetts, won the $1 million contract.

According to Charles M. Herzfeld, the former director of ARPA, Taylor and his colleagues wanted to see if they could link computers and researchers together. The project's military role was much less important. But at the time it was launched, Herzfeld noted, no one knew whether it could be done, so the program, initially funded on $1 million diverted from ballistic-missile defense, was risky.

Taylor became ARPA's computer evangelist, picking up Licklider's mantle and preaching the gospel of distributed interactive computing. In 1968, Taylor and Licklider co-authored a key essay, "The Computer as a Communication Device," which was published in the popular journal Science and Technology. It began with a thunderclap: "In a few years, men will be able to communicate more effectively through a machine than face to face." The article went on to predict everything from global online communities to mood-sensing computer interfaces. It was the first inkling the public ever had about the potential of networked digital computing, and it attracted other researchers to the cause.

A Packet of Data

ARPANET arose from a desire to share information over great distances without the need for dedicated phone connections between each computer on a network. As it turned out, fulfilling this desire would require "packet switching."

Paul Baran, a researcher at the RAND Corporation think tank, first introduced the idea. Baran was instructed to come up with a plan for a computer communications network that could survive nuclear attack and continue functioning. He came up with a process that he called "hot-potato routing," which later became known as packet switching.

"Packets" are small clusters of digital information broken up from larger messages for expediency's sake. To illustrate in more recent terms: an e-mail might be split into numerous electronic packets of information and transmitted almost at random across the labyrinth of the nation's telephone lines. They do not all follow the same route and do not even need to travel in proper sequential order. They are precisely reassembled by a modem at the receiver's end, because each packet contains an identifying "header," revealing which part of the larger message it represents, along with instructions for reconstituting the intended message. As a further safeguard, packets contain mathematical verification schemes that insure data does not get lost in transit. The network on which they travel, meanwhile, consists of computerized switches that automatically forward packets on to their destination. Data packets make computer communications more workable within existing telephone infrastructure by allowing all those packets to flow following paths of least resistance, thereby preventing logjams of digital data over direct, dedicated telephone lines.

Baran's idea was ignored by the military. A 1964 paper outlining his innovation was published, but it was classified and began to collect dust. Fortunately, one place it was collecting dust was in the offices of ARPA, where it was eventually rediscovered. Baran's idea became the key concept that made ARPANET possible. Packet-switched communication remains perhaps the most important legacy handed down to the Internet by ARPANET.

Internet

The Internet (interconnected network) is the global system of interconnected computer networks that use the Internet protocol suite (TCP/IP) to link devices worldwide. It is a *network of networks* that consists of private, public, academic, business, and government networks of local to global scope, linked by a broad array of electronic, wireless, and optical networking technologies. The Internet carries a vast range of information resources and services, such as the inter-linked hypertext documents and applications of the World Wide Web (WWW), electronic mail, telephony, and file sharing.

The origins of the Internet date back to research commissioned by the federal government of the United States in the 1960s to build robust, fault-tolerant communication with computer networks. The primary precursor network, the ARPANET, initially served as a backbone for interconnection of regional academic and military networks in the 1980s. The funding of the National Science Foundation Network as a new backbone in the 1980s, as well as private funding for other commercial extensions, led to worldwide participation in the development of new networking technologies, and the merger of many networks. The linking of commercial networks and enterprises by the early 1990s marked the beginning of the transition to the modern Internet, and generated a sustained exponential growth as generations of institutional, personal, and mobile computers were connected to the network. Although the Internet was widely used by academia since the 1980s, commercialization incorporated its services and technologies into virtually every aspect of modern life.

Most traditional communication media, including telephony, radio, television, paper mail and newspapers are reshaped, redefined, or even bypassed by the Internet, giving birth to new services such as email, Internet telephony, Internet television, online music, digital newspapers, and video streaming websites. Newspaper, book, and other print publishing are adapting to website technology, or are reshaped into blogging, web feeds and online news aggregators. The Internet has enabled and accelerated new forms of personal interactions through instant messaging, Internet forums, and social networking. Online shopping has grown exponentially both for major retailers and small businesses and entrepreneurs, as it enables firms to extend their "brick and mortar" presence to serve a larger market or even sell goods and services entirely online. Business-to-business and financial services on the Internet affect supply chains across entire industries.

The Internet has no single centralized governance in either technological implementation or policies for access and usage; each constituent network sets its own policies. The overreaching definitions of the two principal name spaces in the Internet, the Internet Protocol address (IP address) space and the Domain Name System (DNS), are directed by a maintainer organization, the Internet Corporation for Assigned Names and Numbers (ICANN). The technical underpinning and standardization of the core protocols is an activity of the Internet Engineering Task Force (IETF), a non-profit organization of loosely affiliated international participants that anyone may associate with by contributing technical expertise. In November 2006, the Internet was included on USA Today's list of New Seven Wonders.

Governance

ICANN headquarters in the Playa Vista neighborhood of Los Angeles, California, United States.

The Internet is a global network that comprises many voluntarily interconnected autonomous networks. It operates without a central governing body. The technical underpinning and standardization of the core protocols (IPv4 and IPv6) is an activity of the Internet Engineering Task Force (IETF), a non-profit organization of loosely affiliated international participants that anyone may associate with by contributing technical expertise. To maintain interoperability, the principal name spaces of the Internet are administered by the Internet Corporation for Assigned Names and Numbers (ICANN). ICANN is governed by an international board of directors drawn from across the Internet technical, business, academic, and other non-commercial communities. ICANN coordinates the assignment of unique identifiers for use on the Internet, including domain names, Internet Protocol (IP) addresses, application port numbers in the transport protocols, and many other parameters. Globally unified name spaces are essential for maintaining the global reach of the Internet. This role of ICANN distinguishes it as perhaps the only central coordinating body for the global Internet.

Regional Internet Registries (RIRs) where established for five regions of the world. The African Network Information Center (AfriNIC) for Africa, the American Registry for Internet Numbers (ARIN) for North America, the Asia-Pacific Network Information Centre (APNIC) for Asia and the Pacific region, the Latin American and Caribbean Internet Addresses Registry (LACNIC) for Latin

America and the Caribbean region, and the Réseaux IP Européens – Network Coordination Centre (RIPE NCC) for Europe, the Middle East, and Central Asia were delegated to assign Internet Protocol address blocks and other Internet parameters to local registries, such as Internet service providers, from a designated pool of addresses set aside for each region.

The National Telecommunications and Information Administration, an agency of the United States Department of Commerce, had final approval over changes to the DNS root zone until the IANA stewardship transition on 1 October 2016. The Internet Society (ISOC) was founded in 1992 with a mission to "assure the open development, evolution and use of the Internet for the benefit of all people throughout the world". Its members include individuals (anyone may join) as well as corporations, organizations, governments, and universities. Among other activities ISOC provides an administrative home for a number of less formally organized groups that are involved in developing and managing the Internet, including: the Internet Engineering Task Force (IETF), Internet Architecture Board (IAB), Internet Engineering Steering Group (IESG), Internet Research Task Force (IRTF), and Internet Research Steering Group (IRSG). On 16 November 2005, the United Nations-sponsored World Summit on the Information Society in Tunis established the Internet Governance Forum (IGF) to discuss Internet-related issues.

Infrastructure

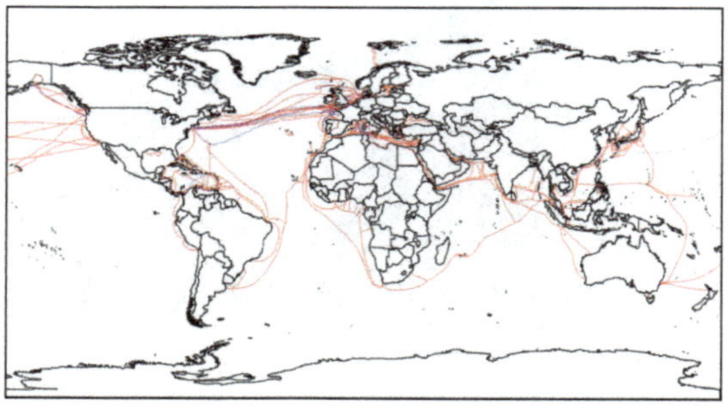

2007 map showing submarine fiberoptic telecommunication cables around the world.

The communications infrastructure of the Internet consists of its hardware components and a system of software layers that control various aspects of the architecture.

Routing and Service Tiers

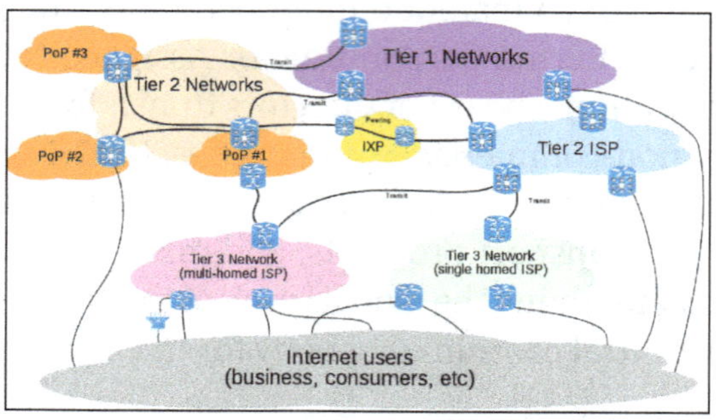

Packet routing across the Internet involves several tiers of Internet service providers.

Internet service providers (ISPs) establish the worldwide connectivity between individual networks at various levels of scope. End-users who only access the Internet when needed to perform a function or obtain information, represent the bottom of the routing hierarchy. At the top of the routing hierarchy are the tier 1 networks, large telecommunication companies that exchange traffic directly with each other via very high speed fibre optic cables and governed by peering agreements. Tier 2 and lower level networks buy Internet transit from other providers to reach at least some parties on the global Internet, though they may also engage in peering. An ISP may use a single upstream provider for connectivity, or implement multihoming to achieve redundancy and load balancing. Internet exchange points are major traffic exchanges with physical connections to multiple ISPs. Large organizations, such as academic institutions, large enterprises, and governments, may perform the same function as ISPs, engaging in peering and purchasing transit on behalf of their internal networks. Research networks tend to interconnect with large subnetworks such as GEANT, GLORIAD, Internet2, and the UK's national research and education network, JANET. Both the Internet IP routing structure and hypertext links of the World Wide Web are examples of scale-free networks. Computers and routers use routing tables in their operating system to direct IP packets to the next-hop router or destination. Routing tables are maintained by manual configuration or automatically by routing protocols. End-nodes typically use a default route that points toward an ISP providing transit, while ISP routers use the Border Gateway Protocol to establish the most efficient routing across the complex connections of the global Internet. An estimated 70 percent of the world's Internet traffic passes through Ashburn, Virginia.

Access

Common methods of Internet access by users include dial-up with a computer modem via telephone circuits, broadband over coaxial cable, fiber optics or copper wires, Wi-Fi, satellite, and cellular telephone technology (e.g. 3G, 4G). The Internet may often be accessed from computers in libraries and Internet cafes. Internet access points exist in many public places such as airport halls and coffee shops. Various terms are used, such as public Internet kiosk, public access terminal, and Web payphone. Many hotels also have public terminals that are usually fee-based. These terminals are widely accessed for various usages, such as ticket booking, bank deposit, or online payment. Wi-Fi provides wireless access to the Internet via local computer networks. Hotspots providing such access include Wi-Fi cafes, where users need to bring their own wireless devices such as a laptop or PDA. These services may be free to all, free to customers only, or fee-based.

Grassroots efforts have led to wireless community networks. Commercial Wi-Fi services that cover large areas are available in many cities, such as New York, London, Vienna, Toronto, San Francisco, Philadelphia, Chicago and Pittsburgh, where the Internet can then be accessed from places such as a park bench. Experiments have also been conducted with proprietary mobile wireless networks like Ricochet, various high-speed data services over cellular networks, and fixed wireless services. Modern smartphones can also access the Internet through the cellular carrier network. For Web browsing, these devices provide applications such as Google Chrome, Safari, and Firefox and a wide variety of other Internet software may be installed from app-stores. Internet usage by mobile and tablet devices exceeded desktop worldwide for the first time in October 2016.

Mobile Communication

The International Telecommunication Union (ITU) estimated that, by the end of 2017, 48% of individual users regularly connect to the Internet, up from 34% in 2012. Mobile Internet connectivity has played an important role in expanding access in recent years especially in Asia and the Pacific and in Africa. The number of unique mobile cellular subscriptions increased from 3.89 billion in 2012 to 4.83 billion in 2016, two-thirds of the world's population, with more than half of subscriptions located in Asia and the Pacific. The number of subscriptions is predicted to rise to 5.69 billion users in 2020. As of 2016, almost 60% of the world's population had access to a 4G broadband cellular network, up from almost 50% in 2015 and 11% in 2012. The limits that users face on accessing information via mobile applications coincide with a broader process of fragmentation of the Internet. Fragmentation restricts access to media content and tends to affect poorest users the most.

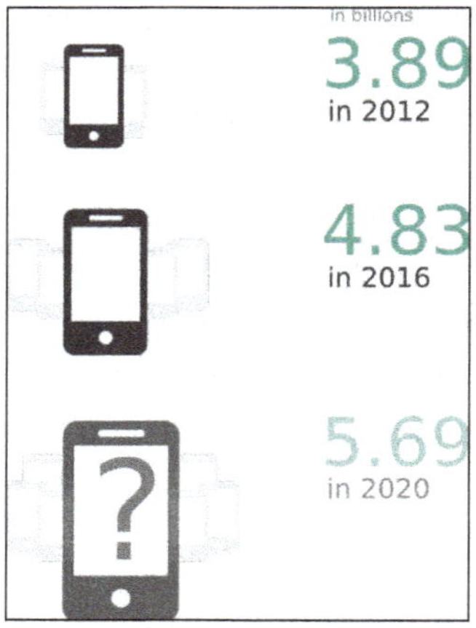

Number of mobile cellular.

Zero-rating, the practice of Internet service providers allowing users free connectivity to access specific content or applications without cost, has offered opportunities to surmount economic hurdles, but has also been accused by its critics as creating a two-tiered Internet. To address the issues with zero-rating, an alternative model has emerged in the concept of 'equal rating' and is being tested in experiments by Mozilla and Orange in Africa. Equal rating prevents prioritization of one type of content and zero-rates all content up to a specified data cap. Chatham House, 15 out of 19 countries researched in Latin America had some kind of hybrid or zero-rated product offered. Some countries in the region had a handful of plans to choose from (across all mobile network operators) while others, such as Colombia, offered as many as 30 pre-paid and 34 post-paid plans.

A study of eight countries in the Global South found that zero-rated data plans exist in every country, although there is a great range in the frequency with which they are offered and actually used in each. The study looked at the top three to five carriers by market share in Bangladesh, Colombia, Ghana, India, Kenya, Nigeria, Peru and Philippines. Across the 181 plans examined, 13 per cent were offering zero-rated services. Another study, covering Ghana, Kenya, Nigeria and South Africa, found Facebook's Free Basics and Wikipedia Zero to be the most commonly zero-rated content.

Protocols

While the hardware components in the Internet infrastructure can often be used to support other software systems, it is the design and the standardization process of the software that characterizes the Internet and provides the foundation for its scalability and success. The responsibility for the architectural design of the Internet software systems has been assumed by the Internet Engineering Task Force (IETF). The IETF conducts standard-setting work groups, open to any individual, about the various aspects of Internet architecture. Resulting contributions and standards are published as Request for Comments (RFC) documents on the IETF web site. The principal methods of networking that enable the Internet are contained in specially designated RFCs that constitute the Internet Standards. Other less rigorous documents are simply informative, experimental, or historical, or document the best current practices (BCP) when implementing Internet technologies.

The Internet standards describe a framework known as the Internet protocol suite, or in short as TCP/IP, based on the first two components. This is a model architecture that divides methods into a layered system of protocols, originally documented in RFC 1122 and RFC 1123. The layers correspond to the environment or scope in which their services operate. At the top is the application layer, space for the application-specific networking methods used in software applications. For example, a web browser program uses the client-server application model and a specific protocol of interaction between servers and clients, while many file-sharing systems use a peer-to-peer paradigm. Below this top layer, the transport layer connects applications on different hosts with a logical channel through the network with appropriate data exchange methods.

Underlying these layers are the networking technologies that interconnect networks at their borders and exchange traffic across them. The Internet layer enables computers to identify and locate each other by Internet Protocol (IP) addresses, and routes their traffic via intermediate (transit) networks. At the bottom of the architecture is the link layer, which provides logical connectivity between hosts on the same network link, such as a local area network (LAN) or a dial-up connection. The model is designed to be independent of the underlying hardware used for the physical connections, which the model does not concern itself with in any detail. Other models have been developed, such as the OSI model, that attempt to be comprehensive in every aspect of communications. While many similarities exist between the models, they are not compatible in the details of description or implementation. Yet, TCP/IP protocols are usually included in the discussion of OSI networking.

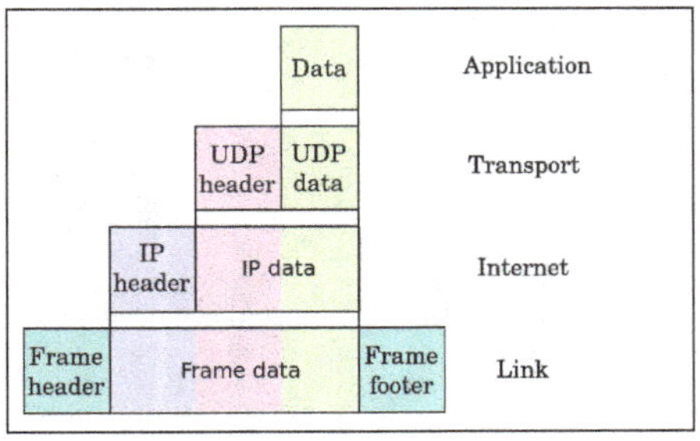

As user data is processed through the protocol stack, each abstraction layer adds encapsulation information at the sending host. Data is transmitted *over the wire* at the link level between hosts and routers. Encapsulation is removed by the receiving host. Intermediate relays update link encapsulation at each hop, and inspect the IP layer for routing purposes.

The most prominent component of the Internet model is the Internet Protocol (IP), which provides addressing systems, including IP addresses, for computers on the network. IP enables internetworking and, in essence, establishes the Internet itself. Internet Protocol Version 4 (IPv4) is the initial version used on the first generation of the Internet and is still in dominant use. It was designed to address up to ≈4.3 billion (10^9) hosts. However, the explosive growth of the Internet has led to IPv4 address exhaustion, which entered its final stage in 2011, when the global address allocation pool was exhausted. A new protocol version, IPv6, was developed in the mid-1990s, which provides vastly larger addressing capabilities and more efficient routing of Internet traffic. IPv6 is currently in growing deployment around the world, since Internet address registries (RIRs) began to urge all resource managers to plan rapid adoption and conversion.

IPv6 is not directly interoperable by design with IPv4. In essence, it establishes a parallel version of the Internet not directly accessible with IPv4 software. Thus, translation facilities must exist for internetworking or nodes must have duplicate networking software for both networks. Essentially all modern computer operating systems support both versions of the Internet Protocol. Network infrastructure, however, has been lagging in this development. Aside from the complex array of physical connections that make up its infrastructure, the Internet is facilitated by bi- or multi-lateral commercial contracts, e.g., peering agreements, and by technical specifications or protocols that describe the exchange of data over the network. Indeed, the Internet is defined by its interconnections and routing policies.

Services

The Internet carries many network services, most prominently the World Wide Web, including social media, electronic mail, mobile applications, multiplayer online games, Internet telephony, file sharing, and streaming media services.

The terms Internet and World Wide Web, or just the Web, are often used interchangeably, but the two terms are not synonymous. The World Wide Web is a primary application program that billions of people use on the Internet, and it has changed their lives immeasurably.

World Wide Web

This NeXT Computer was used by Tim Berners-Lee at CERN and became the world's first Web server.

The World Wide Web is a global collection of documents, images, multimedia, applications, and other resources, logically interrelated by hyperlinks and referenced with Uniform Resource Identifiers (URIs), which provide a global system of named references. URIs symbolically identify services, web servers, databases, and the documents and resources that they can provide. Hypertext Transfer Protocol (HTTP) is the main access protocol of the World Wide Web. Web services also use HTTP for communication between software systems for sharing and exchanging business data and logistics.

World Wide Web browser software, such as Microsoft's Internet Explorer/Edge, Mozilla Firefox, Opera, Apple's Safari, and Google Chrome, lets users navigate from one web page to another via the hyperlinks embedded in the documents. These documents may also contain any combination of computer data, including graphics, sounds, text, video, multimedia and interactive content that runs while the user is interacting with the page. Client-side software can include animations, games, office applications and scientific demonstrations. Through keyword-driven Internet research using search engines like Yahoo!, Bing and Google, users worldwide have easy, instant access to a vast and diverse amount of online information. Compared to printed media, books, encyclopedias and traditional libraries, the World Wide Web has enabled the decentralization of information on a large scale.

The Web has enabled individuals and organizations to publish ideas and information to a potentially large audience online at greatly reduced expense and time delay. Publishing a web page, a blog, or building a website involves little initial cost and many cost-free services are available. However, publishing and maintaining large, professional web sites with attractive, diverse and up-to-date information is still a difficult and expensive proposition. Many individuals and some companies and groups use *web logs* or blogs, which are largely used as easily updatable online diaries. Some commercial organizations encourage staff to communicate advice in their areas of specialization in the hope that visitors will be impressed by the expert knowledge and free information, and be attracted to the corporation as a result.

Advertising on popular web pages can be lucrative, and e-commerce, which is the sale of products and services directly via the Web, continues to grow. Online advertising is a form of marketing and advertising which uses the Internet to deliver promotional marketing messages to consumers. It includes email marketing, search engine marketing (SEM), social media marketing, many types of display advertising (including web banner advertising), and mobile advertising. In 2011, Internet advertising revenues in the United States surpassed those of cable television and nearly exceeded those of broadcast television. Many common online advertising practices are controversial and increasingly subject to regulation.

When the Web developed in the 1990s, a typical web page was stored in completed form on a web server, formatted in HTML, complete for transmission to a web browser in response to a request. Over time, the process of creating and serving web pages has become dynamic, creating a flexible design, layout, and content. Websites are often created using content management software with, initially, very little content. Contributors to these systems, who may be paid staff, members of an organization or the public, fill underlying databases with content using editing pages designed for that purpose while casual visitors view and read this content in HTML form. There may or may not be editorial, approval and security systems built into the process of taking newly entered content and making it available to the target visitors.

Communication

Email is an important communications service available on the Internet. The concept of sending electronic text messages between parties in a way analogous to mailing letters or memos predates the creation of the Internet. Pictures, documents, and other files are sent as email attachments. Emails can be cc-ed to multiple email addresses.

Internet telephony is another common communications service made possible by the creation of the Internet. VoIP stands for Voice-over-Internet Protocol, referring to the protocol that underlies all Internet communication. The idea began in the early 1990s with walkie-talkie-like voice applications for personal computers. In recent years many VoIP systems have become as easy to use and as convenient as a normal telephone. The benefit is that, as the Internet carries the voice traffic, VoIP can be free or cost much less than a traditional telephone call, especially over long distances and especially for those with always-on Internet connections such as cable or ADSL and mobile data. VoIP is maturing into a competitive alternative to traditional telephone service. Interoperability between different providers has improved and the ability to call or receive a call from a traditional telephone is available. Simple, inexpensive VoIP network adapters are available that eliminate the need for a personal computer.

Voice quality can still vary from call to call, but is often equal to and can even exceed that of traditional calls. Remaining problems for VoIP include emergency telephone number dialing and reliability. Currently, a few VoIP providers provide an emergency service, but it is not universally available. Older traditional phones with no "extra features" may be line-powered only and operate during a power failure; VoIP can never do so without a backup power source for the phone equipment and the Internet access devices. VoIP has also become increasingly popular for gaming applications, as a form of communication between players. Popular VoIP clients for gaming include Ventrilo and Teamspeak. Modern video game consoles also offer VoIP chat features.

Data Transfer

File sharing is an example of transferring large amounts of data across the Internet. A computer file can be emailed to customers, colleagues and friends as an attachment. It can be uploaded to a website or File Transfer Protocol (FTP) server for easy download by others. It can be put into a "shared location" or onto a file server for instant use by colleagues. The load of bulk downloads to many users can be eased by the use of "mirror" servers or peer-to-peer networks. In any of these cases, access to the file may be controlled by user authentication, the transit of the file over the Internet may be obscured by encryption, and money may change hands for access to the file. The price can be paid by the remote charging of funds from, for example, a credit card whose details are also passed – usually fully encrypted – across the Internet. The origin and authenticity of the file received may be checked by digital signatures or by MD5 or other message digests. These simple features of the Internet, over a worldwide basis, are changing the production, sale, and distribution of anything that can be reduced to a computer file for transmission. This includes all manner of print publications, software products, news, music, film, video, photography, graphics and the other arts. This in turn has caused seismic shifts in each of the existing industries that previously controlled the production and distribution of these products.

Streaming media is the real-time delivery of digital media for the immediate consumption or enjoyment by end users. Many radio and television broadcasters provide Internet feeds of their live audio and video productions. They may also allow time-shift viewing or listening such as Preview, Classic Clips and Listen Again features. These providers have been joined by a range of pure Internet "broadcasters" who never had on-air licenses. This means that an Internet-connected device, such as a computer or something more specific, can be used to access on-line media in much the same way as was previously possible only with a television or radio receiver. The range of available types of content is much wider, from specialized technical webcasts to on-demand popular multimedia services. Podcasting is a variation on this theme, where – usually audio – material is downloaded and played back on a computer or shifted to a portable media player to be listened to on the move. These techniques using simple equipment allow anybody, with little censorship or licensing control, to broadcast audio-visual material worldwide.

Digital media streaming increases the demand for network bandwidth. For example, standard image quality needs 1 Mbit/s link speed for SD 480p, HD 720p quality requires 2.5 Mbit/s, and the top-of-the-line HDX quality needs 4.5 Mbit/s for 1080p.

Webcams are a low-cost extension of this phenomenon. While some webcams can give full-frame-rate video, the picture either is usually small or updates slowly. Internet users can watch animals around an African waterhole, ships in the Panama Canal, traffic at a local roundabout or monitor their own premises, live and in real time. Video chat rooms and video conferencing are also popular with many uses being found for personal webcams, with and without two-way sound. YouTube was founded on 15 February 2005 and is now the leading website for free streaming video with a vast number of users. It uses a HTML5 based web player by default to stream and show video files. Registered users may upload an unlimited amount of video and build their own personal profile. YouTube claims that its users watch hundreds of millions, and upload hundreds of thousands of videos daily.

Social Impact

The Internet has enabled new forms of social interaction, activities, and social associations. This phenomenon has given rise to the scholarly study of the sociology of the Internet.

Users

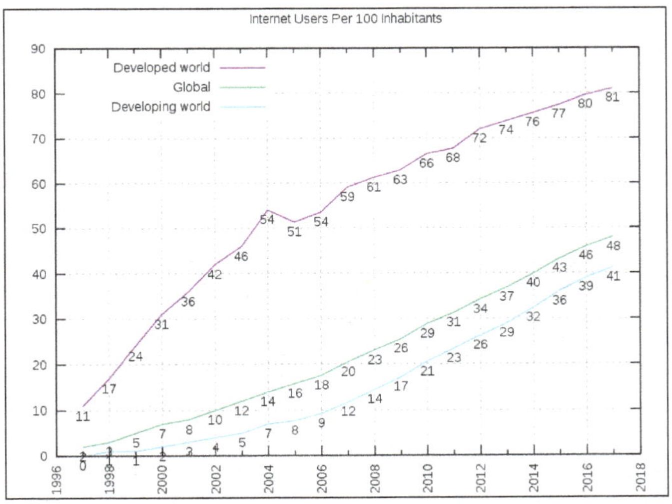

Internet Users per 100 Inhabitants.

Internet usage has grown tremendously. From 2000 to 2009, the number of Internet users globally rose from 394 million to 1.858 billion. By 2010, 22 percent of the world's population had access to computers with 1 billion Google searches every day, 300 million Internet users reading blogs, and 2 billion videos viewed daily on YouTube. In 2014 the world's Internet users surpassed 3 billion or 43.6 percent of world population, but two-thirds of the users came from richest countries, with 78.0 percent of Europe countries population using the Internet, followed by 57.4 percent of the Americas. However, by 2018, this trend had shifted so tremendously that Asia alone accounted for 51% of all Internet users, with 2.2 billion out of the 4.3 billion Internet users in the world coming from that region. The number of China's Internet users surpassed a major milestone in 2018, when the country's Internet regulatory authority, China Internet Network Information Centre, announced that China had 802 million Internet users. By 2019, China was the world's leading country in terms of Internet users, with more than 800 million users, followed closely by India, with some 700 million users, with USA a distant third with 275 million users. However, in terms of penetration, China has a 38.4% penetration rate compared to India's 40% and USA's 80%.

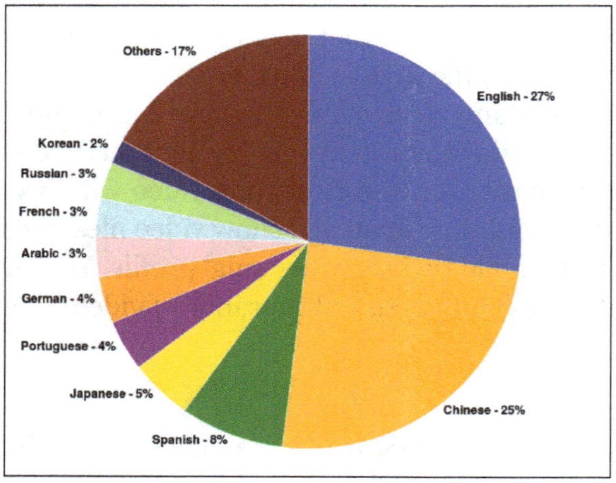

Internet Users by Language.

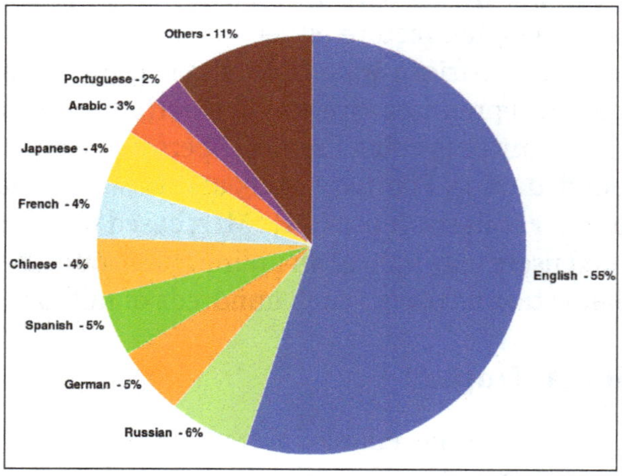

Website Content Languages.

The prevalent language for communication via the Internet has been English. This may be a result of the origin of the Internet, as well as the language's role as a lingua franca. Early computer systems were limited to the characters in the American Standard Code for Information Interchange (ASCII), a subset of the Latin alphabet.

After English (27%), the most requested languages on the World Wide Web are Chinese (25%), Spanish (8%), Japanese (5%), Portuguese and German (4% each), Arabic, French and Russian (3% each), and Korean (2%). By region, 42% of the world's Internet users are based in Asia, 24% in Europe, 14% in North America, 10% in Latin America and the Caribbean taken together, 6% in Africa, 3% in the Middle East and 1% in Australia/Oceania. The Internet's technologies have developed enough in recent years, especially in the use of Unicode, that good facilities are available for development and communication in the world's widely used languages. However, some glitches such as *mojibake* (incorrect display of some languages' characters) still remain.

In an American study in 2005, the percentage of men using the Internet was very slightly ahead of the percentage of women, although this difference reversed in those under 30. Men logged on more often, spent more time online, and were more likely to be broadband users, whereas women tended to make more use of opportunities to communicate (such as email). Men were more likely to use the Internet to pay bills, participate in auctions, and for recreation such as downloading music and videos. Men and women were equally likely to use the Internet for shopping and banking. More recent studies indicate that in 2008, women significantly outnumbered men on most social networking sites, such as Facebook and Myspace, although the ratios varied with age. In addition, women watched more streaming content, whereas men downloaded more. In terms of blogs, men were more likely to blog in the first place; among those who blog, men were more likely to have a professional blog, whereas women were more likely to have a personal blog.

Forecasts predict that 44% of the world's population will be users of the Internet by 2020. Splitting by country, in 2012 Iceland, Norway, Sweden, the Netherlands, and Denmark had the highest Internet penetration by the number of users, with 93% or more of the population with access.

Several neologisms exist that refer to Internet users: Netizen (as in "citizen of the net") refers to those actively involved in improving online communities, the Internet in general or surrounding political affairs and rights such as free speech, Internaut refers to operators or technically highly capable users of the Internet, digital citizen refers to a person using the Internet in order to engage in society, politics, and government participation.

Usage

The Internet allows greater flexibility in working hours and location, especially with the spread of unmetered high-speed connections. The Internet can be accessed almost anywhere by numerous means, including through mobile Internet devices. Mobile phones, datacards, handheld game consoles and cellular routers allow users to connect to the Internet wirelessly. Within the limitations imposed by small screens and other limited facilities of such pocket-sized devices, the services of the Internet, including email and the web, may be available. Service providers may restrict the services offered and mobile data charges may be significantly higher than other access methods.

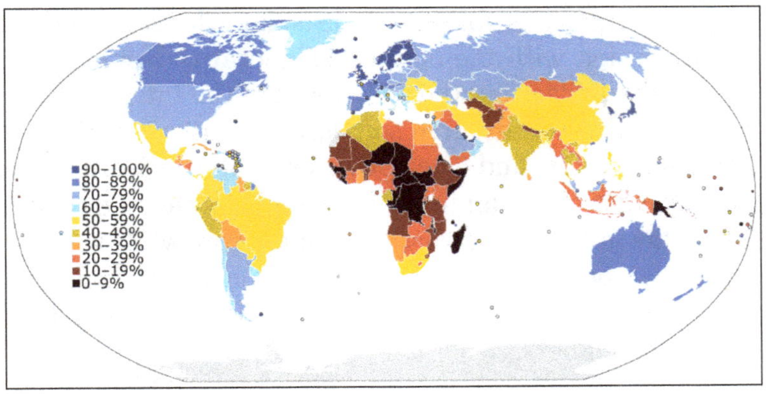

Internet Users in 2015 as a Percentage of a Country's Population.

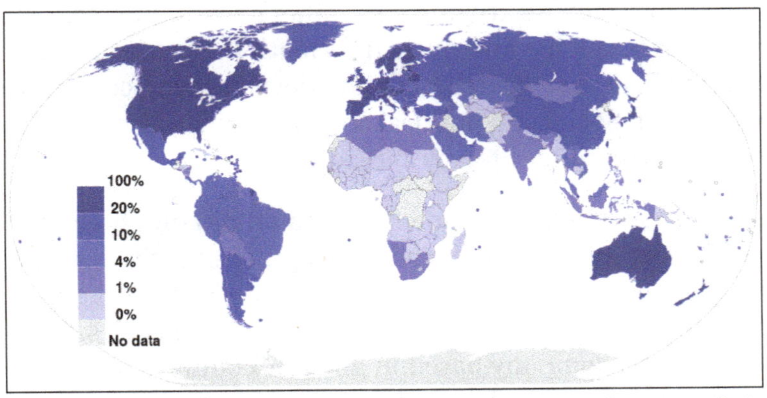

Fixed Broadband Internet Subscriptions in 2012 as a Percentage of a Country's Population.

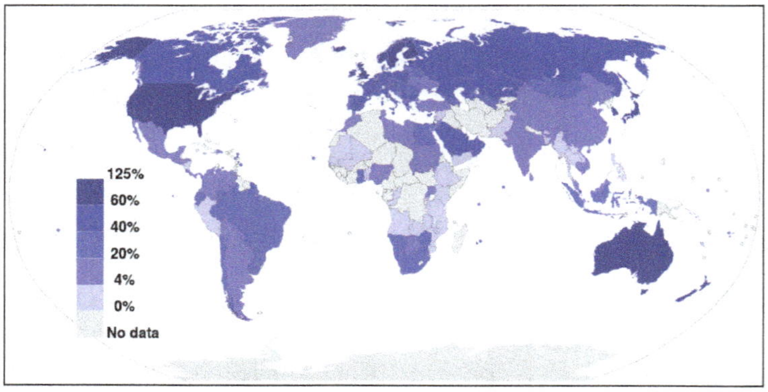

Mobile Broadband Internet Subscriptions in 2012 as a Percentage of a Country's Population.

Educational material at all levels from pre-school to post-doctoral is available from websites. Examples range from CBeebies, through school and high-school revision guides and virtual universities, to access to top-end scholarly literature through the likes of Google Scholar. For distance education, help with homework and other assignments, self-guided learning, whiling away spare time, or just looking up more detail on an interesting fact, it has never been easier for people to access educational information at any level from anywhere. The Internet in general and the World Wide Web in particular are important enablers of both formal and informal education. Further, the Internet allows universities, in particular, researchers from the social and behavioral sciences, to conduct research remotely via virtual laboratories, with profound changes in reach and generalizability of findings as well as in communication between scientists and in the publication of results.

The low cost and nearly instantaneous sharing of ideas, knowledge, and skills have made collaborative work dramatically easier, with the help of collaborative software. Not only can a group cheaply communicate and share ideas but the wide reach of the Internet allows such groups more easily to form. An example of this is the free software movement, which has produced, among other things, Linux, Mozilla Firefox, and OpenOffice.org (later forked into LibreOffice). Internet chat, whether using an IRC chat room, an instant messaging system, or a social networking website, allows colleagues to stay in touch in a very convenient way while working at their computers during the day. Messages can be exchanged even more quickly and conveniently than via email. These systems may allow files to be exchanged, drawings and images to be shared, or voice and video contact between team members.

Content management systems allow collaborating teams to work on shared sets of documents simultaneously without accidentally destroying each other's work. Business and project teams can share calendars as well as documents and other information. Such collaboration occurs in a wide variety of areas including scientific research, software development, conference planning, political activism and creative writing. Social and political collaboration is also becoming more widespread as both Internet access and computer literacy spread.

The Internet allows computer users to remotely access other computers and information stores easily from any access point. Access may be with computer security, i.e. authentication and encryption technologies, depending on the requirements. This is encouraging new ways of working from home, collaboration and information sharing in many industries. An accountant sitting at home can audit the books of a company based in another country, on a server situated in a third country that is remotely maintained by IT specialists in a fourth. These accounts could have been created by home-working bookkeepers, in other remote locations, based on information emailed to them from offices all over the world. Some of these things were possible before the widespread use of the Internet, but the cost of private leased lines would have made many of them infeasible in practice. An office worker away from their desk, perhaps on the other side of the world on a business trip or a holiday, can access their emails, access their data using cloud computing, or open a remote desktop session into their office PC using a secure virtual private network (VPN) connection on the Internet. This can give the worker complete access to all of their normal files and data, including email and other applications, while away from the office. It has been referred to among system administrators as the Virtual Private Nightmare, because it extends the secure perimeter of a corporate network into remote locations and its employees' homes.

Social Networking and Entertainment

Many people use the World Wide Web to access news, weather and sports reports, to plan and book vacations and to pursue their personal interests. People use chat, messaging and email to make and stay in touch with friends worldwide, sometimes in the same way as some previously had pen pals. Social networking websites such as Facebook, Twitter, and Myspace have created new ways to socialize and interact. Users of these sites are able to add a wide variety of information to pages, to pursue common interests, and to connect with others. It is also possible to find existing acquaintances, to allow communication among existing groups of people. Sites like LinkedIn foster commercial and business connections. YouTube and Flickr specialize in users' videos and photographs. While social networking sites were initially for individuals only, today they are widely used by businesses and other organizations to promote their brands, to market to their customers and to encourage posts to "go viral". "Black hat" social media techniques are also employed by some organizations, such as spam accounts and astroturfing.

A risk for both individuals and organizations writing posts (especially public posts) on social networking websites, is that especially foolish or controversial posts occasionally lead to an unexpected and possibly large-scale backlash on social media from other Internet users. This is also a risk in relation to controversial offline behavior, if it is widely made known. The nature of this backlash can range widely from counter-arguments and public mockery, through insults and hate speech, to, in extreme cases, rape and death threats. The online disinhibition effect describes the tendency of many individuals to behave more stridently or offensively online than they would in person. A significant number of feminist women have been the target of various forms of harassment in response to posts they have made on social media, and Twitter in particular has been criticised in the past for not doing enough to aid victims of online abuse.

For organizations, such a backlash can cause overall brand damage, especially if reported by the media. However, this is not always the case, as any brand damage in the eyes of people with an opposing opinion to that presented by the organization could sometimes be outweighed by strengthening the brand in the eyes of others. Furthermore, if an organization or individual gives in to demands that others perceive as wrong-headed, that can then provoke a counter-backlash.

Some websites, such as Reddit, have rules forbidding the posting of personal information of individuals (also known as doxxing), due to concerns about such postings leading to mobs of large numbers of Internet users directing harassment at the specific individuals thereby identified. In particular, the Reddit rule forbidding the posting of personal information is widely understood to imply that all identifying photos and names must be censored in Facebook screenshots posted to Reddit. However, the interpretation of this rule in relation to public Twitter posts is less clear, and in any case, like-minded people online have many other ways they can use to direct each other's attention to public social media posts they disagree with.

Children also face dangers online such as cyberbullying and approaches by sexual predators, who sometimes pose as children themselves. Children may also encounter material which they may find upsetting, or material which their parents consider to be not age-appropriate. Due to naivety, they may also post personal information about themselves online, which could put them or their families at risk unless warned not to do so. Many parents choose to enable Internet filtering, and/or supervise their children's online activities, in an attempt to protect their children from inappropriate material on the Internet. The most popular social networking websites, such as Facebook and Twitter, commonly forbid users under the age of 13. However, these policies are typically trivial to circumvent by registering an account with a false birth date, and a significant number of children aged under 13 join such sites anyway. Social networking sites for younger children, which claim to provide better levels of protection for children, also exist.

The Internet has been a major outlet for leisure activity since its inception, with entertaining social experiments such as MUDs and MOOs being conducted on university servers, and humor-related Usenet groups receiving much traffic. Many Internet forums have sections devoted to games and funny videos. The Internet pornography and online gambling industries have taken advantage of the World Wide Web, and often provide a significant source of advertising revenue for other websites. Although many governments have attempted to restrict both industries' use of the Internet, in general, this has failed to stop their widespread popularity.

Another area of leisure activity on the Internet is multiplayer gaming. This form of recreation creates communities, where people of all ages and origins enjoy the fast-paced world of multiplayer games. These range from MMORPG to first-person shooters, from role-playing video games to online gambling. While online gaming has been around since the 1970s, modern modes of online gaming began with subscription services such as GameSpy and MPlayer. Non-subscribers were limited to certain types of game play or certain games. Many people use the Internet to access and download music, movies and other works for their enjoyment and relaxation. Free and fee-based services exist for all of these activities, using centralized servers and distributed peer-to-peer technologies. Some of these sources exercise more care with respect to the original artists' copyrights than others.

Internet usage has been correlated to users' loneliness. Lonely people tend to use the Internet as an outlet for their feelings and to share their stories with others, such as in the "I am lonely will anyone speak to me" thread.

Cybersectarianism is a new organizational form which involves: "highly dispersed small groups of practitioners that may remain largely anonymous within the larger social context and operate in relative secrecy, while still linked remotely to a larger network of believers who share a set of practices and texts, and often a common devotion to a particular leader. Overseas supporters provide funding and support; domestic practitioners distribute tracts, participate in acts of resistance, and share information on the internal situation with outsiders. Collectively, members and practitioners of such sects construct viable virtual communities of faith, exchanging personal testimonies and engaging in the collective study via email, on-line chat rooms, and web-based message boards." In particular, the British government has raised concerns about the prospect of young British Muslims being indoctrinated into Islamic extremism by material on the Internet, being persuaded to join terrorist groups such as the so-called "Islamic State", and then potentially committing acts of terrorism on returning to Britain after fighting in Syria or Iraq.

Cyberslacking can become a drain on corporate resources; the average UK employee spent 57 minutes a day surfing the Web while at work, according to a 2003 study by Peninsula Business Services. Internet addiction disorder is excessive computer use that interferes with daily life. Nicholas G. Carr believes that Internet use has other effects on individuals, for instance improving skills of scan-reading and interfering with the deep thinking that leads to true creativity.

Electronic Business

Electronic business (*e-business*) encompasses business processes spanning the entire value chain: purchasing, supply chain management, marketing, sales, customer service, and business relationship. E-commerce seeks to add revenue streams using the Internet to build and enhance relationships with clients and partners. According to International Data Corporation, the size of worldwide e-commerce, when global business-to-business and -consumer transactions are combined, equate to $16 trillion for 2013. A report by Oxford Economics adds those two together to estimate the total size of the digital economy at $20.4 trillion, equivalent to roughly 13.8% of global sales.

While much has been written of the economic advantages of Internet-enabled commerce, there is also evidence that some aspects of the Internet such as maps and location-aware services may serve to reinforce economic inequality and the digital divide. Electronic commerce may be responsible for consolidation and the decline of mom-and-pop, brick and mortar businesses resulting in increases in income inequality.

Author Andrew Keen, a long-time critic of the social transformations caused by the Internet, has recently focused on the economic effects of consolidation from Internet businesses. Keen cites a 2013 Institute for Local Self-Reliance report saying brick-and-mortar retailers employ 47 people for every $10 million in sales while Amazon employs only 14. Similarly, the 700-employee room rental start-up Airbnb was valued at $10 billion in 2014, about half as much as Hilton Worldwide, which employs 152,000 people. At that time, transportation network company Uber employed 1,000 full-time employees and was valued at $18.2 billion, about the same valuation as Avis Rent a Car and The Hertz Corporation combined, which together employed almost 60,000 people.

Telecommuting

Telecommuting is the performance within a traditional worker and employer relationship when it is facilitated by tools such as groupware, virtual private networks, conference calling, videoconferencing, and voice over IP (VOIP) so that work may be performed from any location, most conveniently the worker's home. It can be efficient and useful for companies as it allows workers to communicate over long distances, saving significant amounts of travel time and cost. As broadband Internet connections become commonplace, more workers have adequate bandwidth at home to use these tools to link their home to their corporate intranet and internal communication networks.

Collaborative Publishing

Wikis have also been used in the academic community for sharing and dissemination of information across institutional and international boundaries. In those settings, they have been found useful for collaboration on grant writing, strategic planning, departmental documentation, and committee work. The United States Patent and Trademark Office uses a wiki to allow the public to collaborate on finding prior art relevant to examination of pending patent applications. Queens, New York has used a wiki to allow citizens to collaborate on the design and planning of a local park. The English Wikipedia has the largest user base among wikis on the World Wide Web and ranks in the top 10 among all Web sites in terms of traffic.

Politics and Political Revolutions

The Internet has achieved new relevance as a political tool. The presidential campaign of Howard Dean in 2004 in the United States was notable for its success in soliciting donation via the Internet. Many political groups use the Internet to achieve a new method of organizing for carrying out their mission, having given rise to Internet activism, most notably practiced by rebels in the Arab Spring. The New York Times suggested that social media websites, such as Facebook and Twitter, helped people organize the political revolutions in Egypt, by helping activists organize protests, communicate grievances, and disseminate information.

Many have understood the Internet as an extension of the Habermasian notion of the public sphere, observing how network communication technologies provide something like a global civic forum. However, incidents of politically motivated Internet censorship have now been recorded in many countries, including western democracies.

Philanthropy

The spread of low-cost Internet access in developing countries has opened up new possibilities for peer-to-peer charities, which allow individuals to contribute small amounts to charitable projects for other individuals. Websites, such as DonorsChoose and GlobalGiving, allow small-scale donors to direct funds to individual projects of their choice. A popular twist on Internet-based philanthropy is the use of peer-to-peer lending for charitable purposes. Kiva pioneered this concept in 2005, offering the first web-based service to publish individual loan profiles for funding. Kiva raises funds for local intermediary microfinance organizations which post stories and updates on behalf of the borrowers. Lenders can contribute as little as $25 to loans of their choice, and receive their money back as borrowers repay. Kiva falls short of being a pure peer-to-peer charity, in that loans are disbursed before being funded by lenders and borrowers do not communicate with lenders themselves.

However, the recent spread of low-cost Internet access in developing countries has made genuine international person-to-person philanthropy increasingly feasible. In 2009, the US-based nonprofit Zidisha tapped into this trend to offer the first person-to-person microfinance platform to link lenders and borrowers across international borders without intermediaries. Members can fund loans for as little as a dollar, which the borrowers then use to develop business activities that improve their families' incomes while repaying loans to the members with interest. Borrowers access the Internet via public cybercafes, donated laptops in village schools, and even smart phones, then create their own profile pages through which they share photos and information about themselves and their businesses. As they repay their loans, borrowers continue to share updates and dialogue with lenders via their profile pages. This direct web-based connection allows members themselves to take on many of the communication and recording tasks traditionally performed by local organizations, bypassing geographic barriers and dramatically reducing the cost of microfinance services to the entrepreneurs.

Security

Internet resources, hardware, and software components are the target of criminal or malicious attempts to gain unauthorized control to cause interruptions, commit fraud, engage in blackmail or access private information.

Malware

Malware is malicious software used and distributed via the Internet. It includes computer viruses which are copied with the help of humans, computer worms which copy themselves automatically, software for denial of service attacks, ransomware, botnets, and spyware that reports on the activity and typing of users. Usually, these activities constitute cybercrime. Defense theorists have also speculated about the possibilities of cyber warfare using similar methods on a large scale.

Surveillance

The vast majority of computer surveillance involves the monitoring of data and traffic on the Internet. In the United States for example, under the Communications Assistance For Law Enforcement Act, all phone calls and broadband Internet traffic (emails, web traffic, instant messaging, etc.) are required to be available for unimpeded real-time monitoring by Federal law enforcement agencies. Packet capture is the monitoring of data traffic on a computer network. Computers communicate over the Internet by breaking up messages (emails, images, videos, web pages, files, etc.) into small chunks called "packets", which are routed through a network of computers, until they reach their destination, where they are assembled back into a complete "message" again. Packet Capture Appliance intercepts these packets as they are traveling through the network, in order to examine their contents using other programs. A packet capture is an information *gathering* tool, but not an *analysis* tool. That is it gathers "messages" but it does not analyze them and figure out what they mean. Other programs are needed to perform traffic analysis and sift through intercepted data looking for important/useful information. Under the Communications Assistance For Law Enforcement Act all U.S. telecommunications providers are required to install packet sniffing technology to allow Federal law enforcement and intelligence agencies to intercept all of their customers' broadband Internet and voice over Internet protocol (VoIP) traffic.

The large amount of data gathered from packet capturing requires surveillance software that filters and reports relevant information, such as the use of certain words or phrases, the access of certain types of web sites, or communicating via email or chat with certain parties. Agencies, such as the Information Awareness Office, NSA, GCHQ and the FBI, spend billions of dollars per year to develop, purchase, implement, and operate systems for interception and analysis of data. Similar systems are operated by Iranian secret police to identify and suppress dissidents. The required hardware and software was allegedly installed by German Siemens AG and Finnish Nokia.

Censorship

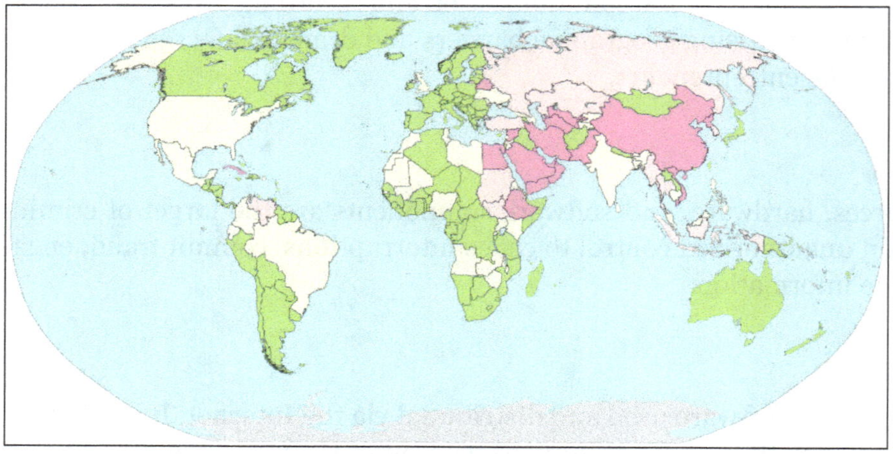

Internet Censorship and Surveillance by Country (2018).

Some governments, such as those of Burma, Iran, North Korea, the Mainland China, Saudi Arabia and the United Arab Emirates restrict access to content on the Internet within their territories, especially to political and religious content, with domain name and keyword filters.

In Norway, Denmark, Finland, and Sweden, major Internet service providers have voluntarily agreed to restrict access to sites listed by authorities. While this list of forbidden resources is supposed to contain only known child pornography sites, the content of the list is secret. Many countries, including the United States, have enacted laws against the possession or distribution of certain material, such as child pornography, via the Internet, but do not mandate filter software. Many free or commercially available software programs, called content-control software are available to users to block offensive websites on individual computers or networks, in order to limit access by children to pornographic material or depiction of violence.

Performance

As the Internet is a heterogeneous network, the physical characteristics, including for example the data transfer rates of connections, vary widely. It exhibits emergent phenomena that depend on its large-scale organization.

Carried Traffic

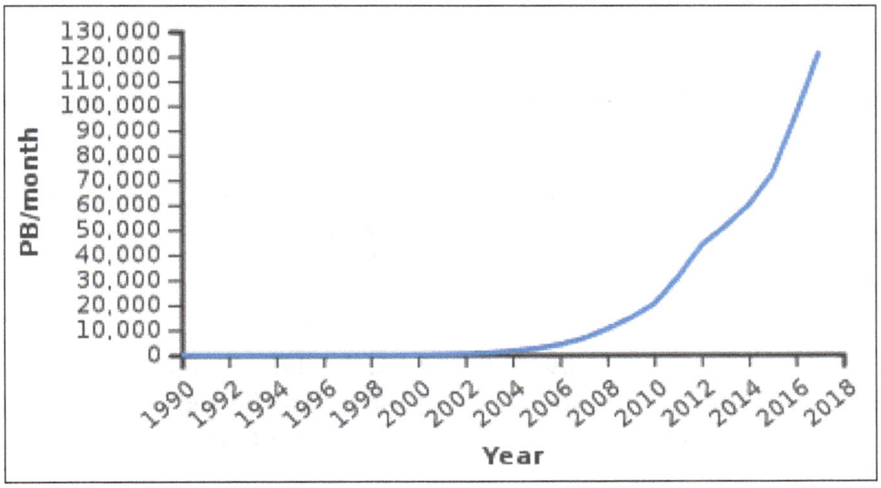

Internet traffic is the flow of data across the Internet, and its size may be measured in terms of bytes.

However, because of the distributed nature of the Internet, there is no single point of measurement for total Internet traffic. Nevertheless, internet traffic data from public peering points can give an indication of Internet volume and growth, but these figures exclude traffic that remains within a single service provider's network as well as traffic that crosses private peering points.

Outages

An Internet blackout or outage can be caused by local signalling interruptions. Disruptions of submarine communications cables may cause blackouts or slowdowns to large areas, such as in the 2008 submarine cable disruption. Less-developed countries are more vulnerable due to a small number of high-capacity links. Land cables are also vulnerable, as in 2011 when a woman digging for scrap metal severed most connectivity for the nation of Armenia. Internet blackouts affecting almost entire countries can be achieved by governments as a form of Internet censorship, as in the blockage of the Internet in Egypt, whereby approximately 93% of networks were without access in 2011 in an attempt to stop mobilization for anti-government protests.

Energy Use

In 2011, researchers estimated the energy used by the Internet to be between 170 and 307 GW, less than two percent of the energy used by humanity. This estimate included the energy needed to build, operate, and periodically replace the estimated 750 million laptops, a billion smart phones and 100 million servers worldwide as well as the energy that routers, cell towers, optical switches, Wi-Fi transmitters and cloud storage devices use when transmitting Internet traffic. According to a study published in 2018, nearly 4% of global CO_2 emission could be attributed to global data transfer and the necessary infrastructure. The study also said that online video streaming alone accounted for 60% of this data transfer and therefore contributed to over 300 million tons of CO_2 emission per year.

Local Area Network

A local area network (LAN) is a computer network that interconnects computers within a limited area such as a residence, school, laboratory, university campus or office building. By contrast, a wide area network (WAN) not only covers a larger geographic distance, but also generally involves leased telecommunication circuits.

Ethernet and Wi-Fi are the two most common technologies in use for local area networks. Historical technologies include ARCNET, Token ring, and AppleTalk.

Cabling

Early Ethernet (10BASE-5 and 10BASE-2) used coaxial cable. Shielded twisted pair was used in IBM's Token Ring LAN implementation. In 1984, StarLAN showed the potential of simple unshielded twisted pair by using Cat3 cable—the same cable used for telephone systems. This led to the development of 10BASE-T (and its successors) and structured cabling which is still the basis of most commercial LANs today.

While optical fiber cable is common for links between network switches, use of fiber to the desktop is rare.

Wireless Media

In a wireless LAN, users have unrestricted movement within the coverage area. Wireless networks have become popular in residences and small businesses, because of their ease of installation. Most wireless LANs use Wi-Fi as it is built into smartphones, tablet computers and laptops. Guests are often offered Internet access via a hotspot service.

Technical Aspects

Network topology describes the layout of interconnections between devices and network segments. At the data link layer and physical layer, a wide variety of LAN topologies have been used, including ring, bus, mesh and star.

Simple LANs generally consist of cabling and one or more switches. A switch can be connected to a router, cable modem, or ADSL modem for Internet access. A LAN can include a wide variety of other network devices such as firewalls, load balancers, and network intrusion detection. Advanced LANs are characterized by their use of redundant links with switches using the spanning tree protocol to prevent loops, their ability to manage differing traffic types via quality of service (QoS), and their ability to segregate traffic with VLANs.

At the higher network layers, protocols such as NetBEUI, IPX/SPX, AppleTalk and others were once common, but the Internet Protocol Suite (TCP/IP) has prevailed as a standard of choice.

LANs can maintain connections with other LANs via leased lines, leased services, or across the Internet using virtual private network technologies. Depending on how the connections are established and secured, and the distance involved, such linked LANs may also be classified as a metropolitan area network (MAN) or a wide area network (WAN).

Wide Area Network

A wide area network (WAN) is a telecommunications network that extends over a large geographical area for the primary purpose of computer networking. Wide area networks are often established with leased telecommunication circuits.

Business, as well as education and government entities use wide area networks to relay data to staff, students, clients, buyers and suppliers from various locations across the world. In essence, this mode of telecommunication allows a business to effectively carry out its daily function regardless of location. The Internet may be considered a WAN.

Similar types of networks are personal area networks (PANs), local area networks (LANs), campus area networks (CANs), or metropolitan area networks (MANs) which are usually limited to a room, building, campus or specific metropolitan area, respectively.

Design Options

The textbook definition of a WAN is a computer network spanning regions, countries, or even the world. However, in terms of the application of computer networking protocols and concepts, it may be best to view WANs as computer networking technologies used to transmit data over long distances, and between different LANs, MANs and other localised computer networking architectures. This distinction stems from the fact that common LAN technologies operating at lower layers of the OSI model (such as the forms of Ethernet or Wi-Fi) are often designed for physically proximal networks, and thus cannot transmit data over tens, hundreds, or even thousands of miles or kilometres.

WANs do not just necessarily connect physically disparate LANs. A CAN, for example, may have a localized backbone of a WAN technology, which connects different LANs within a campus. This could be to facilitate higher bandwidth applications or provide better functionality for users in the CAN.

WANs are used to connect LANs and other types of networks together so that users and computers in one location can communicate with users and computers in other locations. Many WANs are built for one particular organization and are private. Others, built by Internet service providers, provide connections from an organization's LAN to the Internet. WANs are often built using leased lines. At each end of the leased line, a router connects the LAN on one side with a second router within the LAN on the other. Leased lines can be very expensive. Instead of using leased lines, WANs can also be built using less costly circuit switching or packet switching methods. Network protocols including TCP/IP deliver transport and addressing functions. Protocols including Packet over SONET/SDH, Multiprotocol Label Switching (MPLS), Asynchronous Transfer Mode (ATM) and Frame Relay are often used by service providers to deliver the links that are used in WANs. X.25 was an important early WAN protocol, and is often considered to be the "grandfather" of Frame Relay as many of the underlying protocols and functions of X.25 are still in use today (with upgrades) by Frame Relay.

Academic research into wide area networks can be broken down into three areas: mathematical models, network emulation, and network simulation.

Performance improvements are sometimes delivered via wide area file services or WAN optimization.

Private Networks

Of the approximately four billion addresses defined in IPv4, about 18 million addresses in three ranges are reserved for use in private networks. Packets addresses in these ranges are not routable in the public Internet; they are ignored by all public routers. Therefore, private hosts cannot directly communicate with public networks, but require network address translation at a routing gateway for this purpose.

Reserved private IPv4 network ranges				
Name	CIDR block	Address range	Number of addresses	Classful description
24-bit block	10.0.0.0/8	10.0.0.0 – 10.255.255.255	16777216	Single Class A.
20-bit block	172.16.0.0/12	172.16.0.0 – 172.31.255.255	1048576	Contiguous range of 16 Class B blocks.
16-bit block	192.168.0.0/16	192.168.0.0–192.168.255.255	65536	Contiguous range of 256 Class C blocks.

Since two private networks, e.g., two branch offices, cannot directly interoperate via the public Internet, the two networks must be bridged across the Internet via a virtual private network (VPN) or an IP tunnel, which encapsulates packets, including their headers containing the private addresses, in a protocol layer during transmission across the public network. Additionally, encapsulated packets may be encrypted for the transmission across public networks to secure the data.

Connection Technology

Many technologies are available for wide area network links. Examples include circuit-switched telephone lines, radio wave transmission, and optical fiber. New developments in technologies have successively increased transmission rates. In ca. 1960, a 110 bit/s (bits per second) line was normal on the edge of the WAN, while core links of 56 kbit/s to 64 kbit/s were considered fast. As of 2014, households are connected to the Internet with Dial-Up, ADSL, Cable, Wimax, 4G or fiber. The speeds that people can currently use range from 28.8 kbit/s through a 28K modem over a telephone connection to speeds as high as 100 Gbit/s over an Ethernet 100GBaseY connection.

The following communication and networking technologies have been used to implement WANs.

- Asynchronous Transfer Mode
- Cable modem
- Dial-up internet
- Digital subscriber line
- Fiber-optic communication
- Frame Relay
- ISDN
- Leased line
- SD-WAN
- Synchronous optical networking
- X.25

400 Gigabit Ethernet

AT&T conducted trials in 2017 for business use of 400 gigabit Ethernet. Researchers Robert Maher, Alex Alvarado, Domaniç Lavery, and Polina Bayvel of University College London were able to increase networking speeds to 1.125 terabits per second. Christos Santis, graduate student Scott Steger, Amnon Yariv, Martin and Eileen Summerfield developed a new laser that quadruples transfer speeds with fiber optics. If these two technologies were combined, then a transfer speed of up to 4.5 terabits per second could potentially be achieved.

FidoNet

FidoNet is a worldwide computer network that is used for communication between bulletin board systems (BBSes). It uses a store-and-forward system to exchange private (email) and public (forum) messages between the BBSes in the network, as well as other files and protocols in some cases.

The FidoNet system was based on a number of small interacting programs. Only one of these interacted with the BBS system directly and was the only portion that had to be ported to support other BBS software. This greatly eased porting, and FidoNet was one of the few networks that was widely supported by almost all BBS software, as well as a number of non-BBS online services. This modular construction also allowed FidoNet to easily upgrade to new data compression systems, which was important in an era using modem-based communications over telephone links with high long-distance calling charges.

The rapid improvement in modem speeds during the early 1990s, combined with the rapid decrease in price of computer systems and storage, made BBSes increasingly popular. By the mid-1990s there were almost 40,000 FidoNet systems in operation, and it was possible to communicate with millions of users around the world. Only UUCPNET came close in terms of breadth or numbers; FidoNet's user base far surpassed other networks like BITNET.

The broad availability of low-cost Internet connections starting in the mid-1990s lessened the need for FidoNet's store-and-forward system, as any system in the world could be reached for equal cost. Direct dialing into local BBS systems rapidly declined. Although FidoNet has shrunk considerably since the late 1990s, it has remained in use even today despite internet connectivity becoming universally available.

FidoNet Organizational Structure

FidoNet is governed in a hierarchical structure according to FidoNet policy, with designated coordinators at each level to manage the administration of FidoNet nodes and resolve disputes between members. Network coordinators are responsible for managing the individual nodes within their area, usually a city or similar sized area. Regional coordinators are responsible for managing the administration of the network coordinators within their region, typically the size of a state, or small country. Zone coordinators are responsible for managing the administration of all of the regions within their zone. The world is divided into six zones, the coordinators of which elect one of themselves to be the *International Coordinator* of FidoNet.

Technical Structure

FidoNet was historically designed to use modem-based dial-up (POTS) access between bulletin board systems, and much of its policy and structure reflected this.

The FidoNet system officially referred only to transfer of *Netmail*—the individual private messages between people using bulletin boards—including the protocols and standards with which to support it. A netmail message would contain the name of the person sending, the name of the intended recipient, and the respective FidoNet addresses of each. The FidoNet system was responsible for routing the message from one system to the other, with the bulletin board software on each end being responsible for ensuring that only the intended recipient could read it. Due to the hobbyist nature of the network, any privacy between sender and recipient was only the result of politeness from the owners of the FidoNet systems involved in the mail's transfer. It was common, however, for system operators to reserve the right to review the content of mail that passed through their system.

Netmail allowed for the *attachment* of a single file to every message. This led to a series of *piggyback* protocols that built additional features onto FidoNet by passing information back and forth as file attachments. These included the automated distribution of files, and transmission of data for inter-BBS games.

By far the most commonly used of these piggyback protocols was *Echomail*, public discussions similar to Usenet newsgroups in nature. Echomail was supported by a variety of software that collected up new messages from the local BBSes' public forums (the *scanner*), compressed it using ARC or ZIP, attached the resulting archive to a Netmail message, and sent that message to a selected system. On receiving such a message, identified because it was addressed to a particular *user*, the reverse process was used to extract the messages, and a *tosser* put them back into the new system's forums.

Echomail was so popular that for many users, Echomail *was* the FidoNet. Private person-to-person Netmail was relatively rare.

Geographical Structure

FidoNet is politically organized into a tree structure, with different parts of the tree electing their respective coordinators. The FidoNet hierarchy consists of *zones*, *regions*, *networks*, *nodes* and *points* broken down more-or-less geographically.

The highest level is the zone, which is largely continent-based:

- Zone 1 is North America.
- Zone 2 is Europe, Former Soviet Union countries, and Israel.
- Zone 3 is Australasia.
- Zone 4 is Latin America (except Puerto Rico).
- Zone 5 was Africa.
- Zone 6 was Asia, Israel and the Asian parts of Russia, (which are listed in Zone 2). On 26 July 2007 zone 6 was removed, and all remaining nodes were moved to zone 3.

Each zone is broken down into regions, which are broken down into nets, which consist of individual nodes. Zones 7-4095 are used for *othernets*; groupings of nodes which use Fido-compatible software to carry their own independent message areas without being in any way controlled by

FidoNet's political structure. Using un-used zone numbers would ensure that each network would have a unique set of addresses, avoiding potential routing conflicts and ambiguities for systems that belonged to more than one network.

FidoNet Addresses

FidoNet addresses explicitly consist of a *zone* number, a *network* number (or region number), and a *node* number. They are written in the form Zone:Network/Node. The FidoNet structure also allows for semantic designation of region, host, and hub status for particular nodes, but this status is not directly indicated by the main address.

For example, consider a node located in Tulsa, Oklahoma, United States with an assigned node number is 918, located in Zone 1 (North America), Region 19, and Network 170. The full FidoNet address for this system would be 1:170/918. The *region* was used for administrative purposes, and was only part of the address if the node was listed directly underneath the Regional Coordinator, rather than one of the networks that were used to divide the region further.

FidoNet policy requires that each FidoNet system maintain a *nodelist* of every other member system. Information on each node includes the name of the system or BBS, the name of the node operator, the geographic location, the telephone number, and software capabilities. The nodelist is updated weekly, to avoid unwanted calls to nodes that had shut down, with their phone numbers possibly having been reassigned for voice use by the respective telephone company.

To accomplish regular updates, coordinators of each network maintain the list of systems in their local areas. The lists are forwarded back to the International Coordinator via automated systems on a regular basis. The International Coordinator would then compile a new nodelist, and generate the list of changes (nodediff) to be distributed for node operators to apply to their existing nodelist.

Routing of FidoNet Mail

In a theoretical situation, a node would normally forward messages to a *hub*. The hub, acting as a distribution point for mail, might then send the message to the Net Coordinator. From there it may be sent through a Regional Coordinator, or to some other system specifically set up for the function. Mail to other zones might be sent through a Zone Gate.

For example, a FidoNet message might follow the path:

- 1:170/918 (node) to 1:170/900 (hub) to 1:170/0 (net coordinator) to 1:19/0 (region coordinator) to 1:1/0 (zone coordinator). From there, it was distributed 'down stream' to the destination node(s).

Originally there was no specific relationship between network numbers and the regions they reside in. In some areas of FidoNet, most notably in Zone 2, the relationship between region number and network number are entwined. For example, 2:201/329 is in Net 201 which is in Region 20 while 2:2410/330 is in Net 2410 which is in Region 24. Zone 2 also relates the node number to the hub number if the network is large enough to contain any hubs. This effect may be seen in the nodelist by looking at the structure of Net 2410 where node 2:2410/330 is listed under Hub 300. This is not the case in other zones.

In Zone 1, things are much different. Zone 1 was the starting point and when Zones and Regions were formed, the existing nets were divided up regionally with no set formula. The only consideration taken was where they were located geographically in respect to the region's mapped outline. As net numbers got added, the following formula was used:

Region number × 20

Then when some regions started running out of network numbers, the following was also used:

Region number × 200

Region 19, for instance, contains nets 380-399 and 3800-3999 in addition to those that were in Region 19 when it was formed.

Part of the objective behind the formation of local nets was to implement cost reduction plans by which all messages would be sent to one or more hubs or hosts in compressed form (ARC was nominally standard, but PKZIP is universally supported); one toll call could then be made during off-peak hours to exchange entire message-filled archives with an out-of-town uplink for further redistribution.

In practice, the FidoNet structure allows for any node to connect directly to any other, and node operators would sometimes form their own toll-calling arrangements on an ad-hoc basis, allowing for a balance between collective cost saving and timely delivery. For instance, if one node operator in a network offered to make regular toll calls to a particular system elsewhere, other operators might arrange to forward all of their mail destined for the remote system, and those near it, to the local volunteer. Operators within individual networks would sometimes have cost-sharing arrangements, but it was also common for people to volunteer to pay for regular toll calls either out of generosity, or to build their status in the community.

This ad-hoc system was particularly popular with networks that were built on top of FidoNet. Echomail, for instance, often involved relatively large file transfers due to its popularity. If official FidoNet distributors refused to transfer Echomail due to additional toll charges, other node operators would sometimes volunteer. In such cases, Echomail messages would be routed to the volunteers' systems instead.

The FidoNet system was best adapted to an environment in which local telephone service was inexpensive and long-distance calls (or intercity data transfer via packet-switched networks) costly. Therefore, it fared somewhat poorly in Japan, where even local lines are expensive, or in France, where tolls on local calls and competition with Minitel or other data networks limited its growth.

Points

As the number of messages in Echomail grew over time, it became very difficult for users to keep up with the volume while logged into their local BBS. *Points* were introduced to address this, allowing technically savvy users to receive the already compressed and batched Echomail (and Netmail) and read it locally on their own machines.

To do this, the FidoNet addressing scheme was extended with the addition of a final address segment, the point number. For instance, a user on the example system above might be given point number 10, and thus could be sent mail at the address 1:170/918.10.

In real-world use, points are fairly difficult to set up. The FidoNet software typically consisted of a number of small utility programs run by manually edited scripts that required some level of technical ability. Reading and editing the mail required either a "sysop editor" program, or a BBS program to be run locally.

In North America (Zone 1), where local calls are generally free, the benefits of the system were offset by its complexity. Points were used only briefly, and even then only to a limited degree. Dedicated offline mail reader programs such as Blue Wave, Squiggy and Silver Xpress (OPX) were introduced in the mid-1990s, and quickly rendered the point system obsolete. Many of these packages supported the QWK offline mail standard.

In other parts of the world, especially Europe, this was different. In Europe, even local calls are generally metered, so there was a strong incentive to keep the duration of the calls as short as possible. Point software employs standard compression (ZIP, ARJ, etc.) and so keeps the calls down to a few minutes a day at most. In contrast to North America, pointing saw rapid and fairly widespread uptake in Europe.

Many regions distribute a pointlist in parallel with the nodelist. The pointlist segments are maintained by Net- and Region Pointlist Keepers and the Zone Point List Keeper assembles them into the Zone pointlist. At the peak of FidoNet there were over 120,000 points listed in the Zone 2 pointlist. Listing points is on a voluntary basis and not every point is listed, so how many points there really were is anybody's guess. As of June 2006, there are still some 50,000 listed points. Most of them are in Russia and Ukraine.

Technical Specifications

FidoNet contained several technical specifications for compatibility between systems. The most basic of all is *FTS-0001*, with which all FidoNet systems are required to comply as a minimal requirement. FTS-0001 defined:

- Handshaking: The protocols used by mailer software to identify each other and exchange meta information about the session.
- Transfer protocol (XMODEM): The protocols to be used for transferring files containing FidoNet mail between systems.
- Message format: The standard format for FidoNet messages during the time which they were exchanged between systems.

Other specifications that were commonly used provided for *echomail*, different transfer protocols and handshake methods (*e.g.: Yoohoo/Yoohoo2u2, EMSI*), file compression, nodelist format, transfer over reliable connections such as the Internet (Binkp), and other aspects.

Zone Mail Hour

Since computer bulletin boards historically used the same telephone lines for transferring mail as were used for dial-in human users of the BBS, FidoNet policy dictates that at least one designated line of each FidoNet node must be available for accepting mail from other FidoNet nodes during a particular hour of each day.

Zone Mail Hour, as it was named, varies depending on the geographic location of the node, and was designated to occur during the early morning. The exact hour varies depending on the time zone, and any node with only one telephone line is required to reject human callers. In practice, particularly in later times, most FidoNet systems tend to accept mail at any time of day when the phone line is not busy, usually during night.

FidoNet Deployments

Most FidoNet deployments were designed in a modular fashion. A typical deployment would involve several applications that would communicate through shared files and directories, and switch between each other through carefully designed scripts or batch files. However, monolithic software that encompassed all required functions in one package is available, such as D'Bridge. Such software eliminated the need for custom batch files and is tightly integrated in operation. The preference of deployment was that of the operator and there were both pros and cons of running in either fashion.

Arguably the most important piece of software on a DOS-based Fido system was the FOSSIL driver, which was a small device driver which provided a standard way for the Fido software to talk to the modem. This driver needed to be loaded before any Fido software would work. An efficient FOSSIL driver meant faster, more reliable connections.

Mailer software was responsible for transferring files and messages between systems, as well as passing control to other applications, such as the BBS software, at appropriate times. The mailer would initially answer the phone and, if necessary, deal with incoming mail via FidoNet transfer protocols. If the mailer answered the phone and a human caller was detected rather than other mailer software, the mailer would exit, and pass control to the BBS software, which would then initialise for interaction with the user. When outgoing mail was waiting on the local system, the mailer software would attempt to send it from time to time by dialing and connecting to other systems who would accept and route the mail further. Due to the costs of toll calls which often varied between peak and off-peak times, mailer software would usually allow its operator to configure the optimal times in which to attempt to send mail to other systems.

BBS software was used to interact with human callers to the system. BBS software would allow dial-in users to use the system's message bases and write mail to others, locally or on other BBSes. Mail directed to other BBSes would later be routed and sent by the mailer, usually after the user had finished using the system. Many BBSes also allowed users to exchange files, play games, and interact with other users in a variety of ways (i.e.: node to node chat).

A scanner/tosser application, such as FastEcho, FMail, TosScan and Squish, would normally be invoked when a BBS user had entered a new FidoNet message that needed to be sent, or when a mailer had received new mail to be imported into the local messages bases. This application would be responsible for handling the packaging of incoming and outgoing mail, moving it between the local system's message bases and the mailer's inbound and outbound directories. The scanner/tosser application would generally be responsible for basic routing information, determining which systems to forward mail to.

In later times, message readers or editors that were independent of BBS software were also developed. Often the System Operator of a particular BBS would use a devoted message reader, rather than the BBS software itself, to read and write FidoNet and related messages. One of the most popular editors in 2008 was GoldED+. In some cases FidoNet nodes, or more often FidoNet points, had no public bulletin board attached, and existed only for the transfer of mail for the benefit of the node's operator. Most nodes in 2009 had no BBS access, but only points, if anything.

The original *Fido BBS* software, and some other FidoNet-supporting software from the 1980s, is no longer functional on modern systems. This is for several reasons, including problems related to the Y2K bug. In some cases, the original authors have left the BBS or shareware community, and the software, much of which was closed source, has been rendered abandonware.

Several DOS based legacy FidoNet Mailers such as FrontDoor, Intermail, MainDoor and D'Bridge from the early 1990s can still be run today under Windows without a modem, by using the freeware NetFoss Telnet FOSSIL driver, and by using a Virtual Modem such as NetSerial. This allows the mailer to *dial* an IP address or hostname via Telnet, rather than dialing a real POTS phone number. There are similar solutions for Linux such as MODEMU (modem emulator) which has limited success when combined with DOSEMU (DOS emulator). Mail Tossers such as FastEcho and FMail are still used today under both Windows and Linux/DOSEMU.

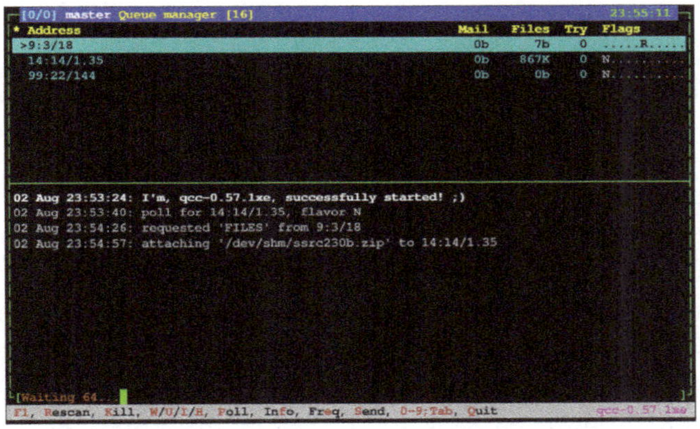

File queue in qcc, the ncurses UI for qico. The addresses are made-up.

There are several modern Windows based FidoNet Mailers available today with source code, including Argus, Radius, and Taurus. MainDoor is another Windows based Fidonet mailer, which also can be run using either a modem or directly over TCP/IP. Two popular free and open source software FidoNet mailers for Unix-like systems are the binkd (cross-platform, IP-only, uses the binkp protocol) and qico (supports modem communication as well as the IP protocol of ifcico and binkp).

On the *hardware* side, Fido systems were usually well-equipped machines, for their day, with quick CPUs, high-speed modems and 16550 UARTs, which were at the time an upgrade. As a Fidonet system was usually a BBS, it needed to quickly process any new mail events before returning to its 'waiting for call' state. In addition, the BBS itself usually necessitated lots of storage space. Finally, a FidoNet system usually had at least one dedicated phoneline. Consequently, operating a Fidonet system often required significant financial investment, a cost usually met by the owner of the system.

FidoNet Availability

While the use of FidoNet has dropped dramatically compared with its use up to the mid-1990s, it is still used in many countries and especially Russia and former republics of the USSR. Some BBSes, including those that are now available for users with Internet connections via telnet, also retain their FidoNet netmail and echomail feeds.

Some of FidoNet's echomail conferences are available via gateways with the Usenet news hierarchy using software like UFGate. There are also mail gates for exchanging messages between Internet and FidoNet. Widespread net abuse and e-mail spam on the Internet side has caused some gateways (such as the former 1:1/31 IEEE fidonet.org gateway) to become unusable or cease operation entirely.

INTEGRATED SERVICES DIGITAL NETWORK

Integrated Services Digital Network (ISDN) is a set of communication standards for simultaneous digital transmission of voice, video, data, and other network services over the traditional circuits of the public switched telephone network. Prior to ISDN, the telephone system was viewed as a way to transport voice, with some special services available for data. The key feature of ISDN is that it integrates speech and data on the same lines, adding features that were not available in the classic telephone system. The ISDN standards define several kinds of access interfaces, such as Basic Rate Interface (BRI), Primary Rate Interface (PRI), Narrowband ISDN (N-ISDN), and Broadband ISDN (B-ISDN).

ISDN is a circuit-switched telephone network system, which also provides access to packet switched networks, designed to allow digital transmission of voice and data over ordinary telephone copper wires, resulting in potentially better voice quality than an analog phone can provide. It offers circuit-switched connections (for either voice or data), and packet-switched connections (for data), in increments of 64 kilobit/s. In some countries, ISDN found major market application for Internet access, in which ISDN typically provides a maximum of 128 kbit/s bandwidth in both upstream and downstream directions. Channel bonding can achieve a greater data rate; typically the ISDN B-channels of three or four BRIs (six to eight 64 kbit/s channels) are bonded.

ISDN is employed as the network, data-link and physical layers in the context of the OSI model. In common use, ISDN is often limited to usage to Q.931 and related protocols, which are a set of signaling protocols establishing and breaking circuit-switched connections, and for advanced calling features for the user. They were introduced in 1986.

In a videoconference, ISDN provides simultaneous voice, video, and text transmission between individual desktop videoconferencing systems and group (room) videoconferencing systems.

ISDN Elements

Integrated services refers to ISDN's ability to deliver at minimum two simultaneous connections, in any combination of data, voice, video, and fax, over a single line. Multiple devices can be attached to the line, and used as needed. That means an ISDN line can take care of what were expected to be most people's complete communications needs (apart from broadband Internet access and entertainment television) at a much higher transmission rate, without forcing the purchase of multiple analog phone lines. It also refers to integrated switching and transmission in that telephone switching and carrier wave transmission are integrated rather than separate as in earlier technology.

Basic Rate Interface

The entry level interface to ISDN is the Basic Rate Interface (BRI), a 128 kbit/s service delivered over a pair of standard telephone copper wires. The 144 kbit/s overall payload rate is divided into two 64 kbit/s bearer channels ('B' channels) and one 16 kbit/s signaling channel ('D' channel or data channel). This is sometimes referred to as 2B+D.

The interface specifies the following network interfaces:

- The U interface is a two-wire interface between the exchange and a network terminating unit, which is usually the demarcation point in non-North American networks.
- The T interface is a serial interface between a computing device and a terminal adapter, which is the digital equivalent of a modem.
- The S interface is a four-wire bus that ISDN consumer devices plug into; the S & T reference points are commonly implemented as a single interface labeled 'S/T' on a Network termination 1 (NT1).
- The R interface defines the point between a non-ISDN device and a terminal adapter (TA) which provides translation to and from such a device.

BRI-ISDN is very popular in Europe but is much less common in North America. It is also common in Japan — where it is known as INS64.

Primary Rate Interface

The other ISDN access available is the Primary Rate Interface (PRI), which is carried over T-carrier (T1) with 24 time slots (channels) in North America, and over E-carrier (E1) with 32 channels in most other countries. Each channel provides transmission at a 64 kbit/s data rate.

With the E1 carrier, the available channels are divided into 30 bearer (*B*) channels, one data (*D*) channel, and one timing and alarm channel. This scheme is often referred to as 30B+2D.

In North America, PRI service is delivered via T1 carriers with only one data channel, often referred to as 23B+D, and a total data rate of 1544 kbit/s. Non-Facility Associated Signalling (NFAS) allows two or more PRI circuits to be controlled by a single D channel, which is sometimes called *23B+D + n*24B*. D-channel backup allows for a second D channel in case the primary fails. NFAS is commonly used on a Digital Signal 3 (DS3/T3).

PRI-ISDN is popular throughout the world, especially for connecting private branch exchanges to the public switched telephone network (PSTN).

Even though many network professionals use the term *ISDN* to refer to the lower-bandwidth BRI circuit, in North America BRI is relatively uncommon whilst PRI circuits serving PBXs are commonplace.

Bearer Channel

The bearer channel (B) is a standard 64 kbit/s voice channel of 8 bits sampled at 8 kHz with G.711 encoding. B-channels can also be used to carry data, since they are nothing more than digital channels.

Each one of these channels is known as a DS0.

Most B channels can carry a 64 kbit/s signal, but some were limited to 56K because they traveled over RBS lines. This was commonplace in the 20th century, but has since become less so.

X.25

X.25 can be carried over the B or D channels of a BRI line, and over the B channels of a PRI line. X.25 over the D channel is used at many point-of-sale (credit card) terminals because it eliminates the modem setup, and because it connects to the central system over a B channel, thereby eliminating the need for modems and making much better use of the central system's telephone lines.

X.25 was also part of an ISDN protocol called "Always On/Dynamic ISDN", or AO/DI. This allowed a user to have a constant multi-link PPP connection to the internet over X.25 on the D channel, and brought up one or two B channels as needed.

Frame Relay

In theory, Frame Relay can operate over the D channel of BRIs and PRIs, but it is seldom, if ever, used.

Consumer and Industry Perspectives

There is a second viewpoint: that of the telephone industry, where ISDN is a core technology. A telephone network can be thought of as a collection of wires strung between switching systems. The common electrical specification for the signals on these wires is T1 or E1. Between telephone company switches, the signaling is performed via SS7. Normally, a PBX is connected via a T1 with robbed bit signaling to indicate on-hook or off-hook conditions and MF and DTMF tones to encode the destination number. ISDN is much better because messages can be sent much more quickly than by trying to encode numbers as long (100 ms per digit) tone sequences. This results in faster call setup times. Also, a greater number of features are available and fraud is reduced.

ISDN is also used as a smart-network technology intended to add new services to the public switched telephone network (PSTN) by giving users direct access to end-to-end circuit-switched digital services and as a backup or failsafe circuit solution for critical use data circuits.

ISDN and Broadcast Industry

ISDN is used heavily by the broadcast industry as a reliable way of switching low-latency, high-quality, long-distance audio circuits. In conjunction with an appropriate codec using MPEG or various manufacturers' proprietary algorithms, an ISDN BRI can be used to send stereo bi-directional audio coded at 128 kbit/s with 20 Hz – 20 kHz audio bandwidth, although commonly the G.722

algorithm is used with a single 64 kbit/s B channel to send much lower latency mono audio at the expense of audio quality. Where very high quality audio is required multiple ISDN BRIs can be used in parallel to provide a higher bandwidth circuit switched connection. BBC Radio 3 commonly makes use of three ISDN BRIs to carry 320 kbit/s audio stream for live outside broadcasts. ISDN BRI services are used to link remote studios, sports grounds and outside broadcasts into the main broadcast studio. ISDN via satellite is used by field reporters around the world. It is also common to use ISDN for the return audio links to remote satellite broadcast vehicles.

In many countries, such as the UK and Australia, ISDN has displaced the older technology of equalised analogue landlines, with these circuits being phased out by telecommunications providers. Use of IP-based streaming codecs such as Comrex ACCESS and ipDTL is becoming more widespread in the broadcast sector, using broadband internet to connect remote studios.

Global Usage

United States and Canada

ISDN-BRI never gained popularity as a general use telephone access technology in Canada and the US, and remains a niche product. The service was seen as "a solution in search of a problem", and the extensive array of options and features were difficult for customers to understand and use. ISDN has long been known by derogatory backronyms highlighting these issues, such as It Still Does Nothing, Innovations Subscribers Don't Need, and I Still Don't kNow.

Once the term "broadband Internet access" came to be associated with data rates incoming to the customer at 256 kbit/s or more, and alternatives like ADSL grew in popularity, the consumer market for BRI did not develop. Its only remaining advantage is that, while ADSL has a functional distance limitation and can use ADSL loop extenders, BRI has a greater limit and can use repeaters. As such, BRI may be acceptable for customers who are too remote for ADSL. Widespread use of BRI is further stymied by some small North American CLECs such as CenturyTel having given up on it and not providing Internet access using it. However, AT&T in most states (especially the former SBC/SWB territory) will still install an ISDN BRI line anywhere a normal analog line can be placed and the monthly charge is roughly $55.

ISDN-BRI is currently primarily used in industries with specialized and very specific needs. High-end videoconferencing hardware can bond up to 8 B-channels together (using a BRI circuit for every 2 channels) to provide digital, circuit-switched video connections to almost anywhere in the world. This is very expensive, and is being replaced by IP-based conferencing, but where cost concern is less of an issue than predictable quality and where a QoS-enabled IP does not exist, BRI is the preferred choice.

Most modern non-VoIP PBXs use ISDN-PRI circuits. These are connected via T1 lines with the central office switch, replacing older analog two-way and direct inward dialing (DID) trunks. PRI is capable of delivering Calling Line Identification (CLID) in both directions so that the telephone number of an extension, rather than a company's main number, can be sent. It is still commonly used in recording studios, when a voice-over actor is in one studio (possibly telecommuting from home), but the director and producer are in a studio at another location. The ISDN protocol delivers channelized, not-over-the-Internet service, powerful call setup and routing features, faster setup and tear down, superior audio fidelity as compared to POTS (plain old telephone service), lower delay and, at higher densities, lower cost.

In 2013, Verizon announced it would no longer take orders for ISDN service in the Northeastern United States.

Australia

Telstra provides the business customer with the ISDN services. There are five types of ISDN services which are ISDN2, ISDN2 Enhanced, ISDN10, ISDN20 and ISDN30. Telstra changed the minimum monthly charge for voice and data calls. In general, there are two group of ISDN service types; The Basic Rate services – ISDN 2 or ISDN 2 Enhanced. Another group of types are the Primary Rate services, ISDN 10/20/30 . Telstra announced that the new sales of ISDN product would be unavailable as of 31 January 2018. The final exit date of ISDN service and migration to the new service would be confirmed by 2022.

India

Bharat Sanchar Nigam Limited, Reliance Communications and Bharti Airtel are the largest communication service providers, and offer both ISDN BRI and PRI services across the country. Reliance Communications and Bharti Airtel uses the DLC technology for providing these services. With the introduction of broadband technology, the load on bandwidth is being absorbed by ADSL. ISDN continues to be an important backup network for point-to-point leased line customers such as banks, Eseva Centers, Life Insurance Corporation of India, and SBI ATMs.

Japan

On April 19, 1988, Japanese telecommunications company NTT began offering nationwide ISDN services trademarked INS Net 64, and INS Net 1500, a fruition of NTT's independent research and trial from the 1970s of what it referred to the INS (Information Network System).

Previously, in April 1985, Japanese digital telephone exchange hardware made by Fujitsu was used to experimentally deploy the world's first I interface ISDN. The I interface, unlike the older and incompatible Y interface, is what modern ISDN services use today.

Since 2000, NTT's ISDN offering have been known as FLET's ISDN, incorporating the "FLET's" brand that NTT uses for all of its ISP offerings.

In Japan, the number of ISDN subscribers dwindled as alternative technologies such as ADSL, cable Internet access, and fiber to the home gained greater popularity. On November 2, 2010, NTT announced plans to migrate their backend from PSTN to the IP network from around 2020 to around 2025. For this migration, ISDN services will be retired, and fiber optic services are recommended as an alternative.

United Kingdom

In the United Kingdom, British Telecom (BT) provides ISDN2e (BRI) as well as ISDN30 (PRI). Until April 2006, they also offered services named Home Highway and Business Highway, which were BRI ISDN-based services that offered integrated analogue connectivity as well as ISDN. Later versions of the Highway products also included built-in USB sockets for direct computer access. Home Highway was bought by many home users, usually for Internet connection, although not as fast as ADSL, because it was available before ADSL and in places where ADSL does not reach.

In early 2015, BT announced their intention to retire the UK's ISDN infrastructure by 2025.

France

France Telecom offers ISDN services under their product name Numeris (2 B+D), of which a professional Duo and home Itoo version is available. ISDN is generally known as RNIS in France and has widespread availability. The introduction of ADSL is reducing ISDN use for data transfer and Internet access, although it is still common in more rural and outlying areas, and for applications such as business voice and point-of-sale terminals.

Germany

German stamp.

In Germany, ISDN was very popular with an installed base of 25 million channels (29% of all subscriber lines in Germany as of 2003 and 20% of all ISDN channels worldwide). Due to the success of ISDN, the number of installed analog lines was decreasing. Deutsche Telekom (DTAG) offered both BRI and PRI. Competing phone companies often offered ISDN only and no analog lines. However, these operators generally offered free hardware that also allows the use of POTS equipment, such as NTBAs with integrated terminal adapters. Because of the widespread availability of ADSL services, ISDN was primarily used for voice and fax traffic.

Until 2007 ISDN (BRI) and ADSL/VDSL were often bundled on the same line, mainly because the combination of DSL with an analog line had no cost advantage over a combined ISDN-DSL line. This practice turned into an issue for the operators when vendors of ISDN technology stopped manufacturing it and spare parts became hard to come by. Since then phone companies started introducing cheaper xDSL-only products using VoIP for telephony, also in an effort to reduce their costs by operating separate data & voice networks.

Since approximately 2010, most german operators are offering more and more VoIP on top of DSL lines and ceased offering ISDN lines. As from 2018 on, new ISDN lines are not available anymore in Germany, existing ISDN lines are phased out from 2016 onwards and existing customers are encouraged to move to DSL-based VoIP products. Deutsche Telekom expected to complete this phase-out by 2018 but postponed the date to 2020, other provides like Vodafone estimate to have their phase-out completed by 2022.

Greece

OTE, the incumbent telecommunications operator, offers ISDN BRI (BRA) services in Greece. Following the launch of ADSL in 2003, the importance of ISDN for data transfer began to decrease and is today limited to niche business applications with point-to-point requirements.

International Deployment

A study of the German Department of Science shows the following spread of ISDN-channels per 1,000 inhabitants in the year 2005:

- Norway 401
- Denmark 339
- Germany 333
- Switzerland 331
- Japan 240
- United Kingdom 160
- Finland 160
- Sweden 135
- Italy 105
- France 85
- Spain 58
- United States 47

Configurations

In ISDN, there are two types of channels, *B* (for "bearer") and *D* (for "data"). *B channels* are used for data (which may include voice), and *D channels* are intended for signaling and control (but can also be used for data).

There are two ISDN implementations. Basic Rate Interface (BRI), also called basic rate access (BRA) — consists of two B channels, each with bandwidth of 64 kbit/s, and one D channel with a bandwidth of 16 kbit/s. Together these three channels can be designated as 2B+D. Primary Rate Interface (PRI), also called primary rate access (PRA) in Europe — contains a greater number of B channels and a D channel with a bandwidth of 64 kbit/s. The number of B channels for PRI varies according to the nation: in North America and Japan it is 23B+1D, with an aggregate bit rate of 1.544 Mbit/s (T1); in Europe, India and Australia it is 30B+2D, with an aggregate bit rate of 2.048 Mbit/s (E1). Broadband Integrated Services Digital Network (BISDN) is another ISDN implementation and it is able to manage different types of services at the same time. It is primarily used within network backbones and employs ATM.

Another alternative ISDN configuration can be used in which the B channels of an ISDN BRI line are bonded to provide a total duplex bandwidth of 128 kbit/s. This precludes use of the line for voice calls while the internet connection is in use. The B channels of several BRIs can be bonded, a typical use is a 384K videoconferencing channel.

Using bipolar with eight-zero substitution encoding technique, call data is transmitted over the data (B) channels, with the signaling (D) channels used for call setup and management. Once a call is set up, there is a simple 64 kbit/s synchronous bidirectional data channel (actually implemented as two simplex channels, one in each direction) between the end parties, lasting until the call is terminated. There can be as many calls as there are bearer channels, to the same or different end-points. Bearer channels may also be multiplexed into what may be considered single, higher-bandwidth channels via a process called B channel BONDING, or via use of Multi-Link PPP "bundling" or by using an H0, H11, or H12 channel on a PRI.

The D channel can also be used for sending and receiving X.25 data packets, and connection to X.25 packet network, this is specified in X.31. In practice, X.31 was only commercially implemented in the UK, France, Japan and Germany.

Reference Points

A set of *reference points* are defined in the ISDN standard to refer to certain points between the telco and the end user ISDN equipment.

- R – defines the point between a non-ISDN terminal equipment 2 (TE2) device and a *terminal adapter* (TA) which provides translation to and from such a device
- S – defines the point between the ISDN terminal equipment 1 (TE1) or TA and a *Network Termination Type 2* (NT2) device
- T – defines the point between the NT2 and network termination 1 (NT1) devices.

Most NT-1 devices can perform the functions of the NT2 as well, and so the S and T reference points are generally collapsed into the S/T reference point.

In North America, the NT1 device is considered customer premises equipment (CPE) and must be maintained by the customer, thus, the U interface is provided to the customer. In other locations, the NT1 device is maintained by the telco, and the S/T interface is provided to the customer. In India, service providers provide U interface and an NT1 may be supplied by Service provider as part of service offering.

Types of Communications

Among the kinds of data that can be moved over the 64 kbit/s channels are pulse-code modulated voice calls, providing access to the traditional voice PSTN. This information can be passed between the network and the user end-point at call set-up time. In North America, ISDN is now used mostly as an alternative to analog connections, most commonly for Internet access. Some of the services envisioned as being delivered over ISDN are now delivered over the Internet instead. In Europe, and in Germany in particular, ISDN has been successfully marketed as a phone with features, as opposed to a POTS phone with few or no features. Meanwhile, features that were first available with ISDN (such as Three-Way Calling, Call Forwarding, Caller ID, etc.) are now commonly available for ordinary analog phones as well, eliminating this advantage of ISDN. Another advantage of ISDN was the possibility of multiple simultaneous calls (one call per B channel), e.g. for big families, but with the increased popularity and reduced prices of mobile telephony this has become less interesting as well, making ISDN unappealing to the private customer. However, ISDN is typically more reliable than POTS, and has a significantly faster call setup time compared with POTS, and IP connections over ISDN typically have some 30–35ms round trip time, as opposed to 120–180ms (both measured with otherwise unused lines) over 56k or V.34/V.92 modems, making ISDN more reliable and more efficient for telecommuters.

Where an analog connection requires a modem, an ISDN connection requires a terminal adapter (TA). The function of an ISDN terminal adapter is often delivered in the form of a PC card with an

S/T interface, and single-chip solutions seem to exist, considering the plethora of combined ISDN- and ADSL-routers.

ISDN is commonly used in radio broadcasting. Since ISDN provides a high quality connection this assists in delivering good quality audio for transmission in radio. Most radio studios are equipped with ISDN lines as their main form of communication with other studios or standard phone lines. Equipment made by companies such as Telos/Omnia (the popular Zephyr codec), Comrex, Tieline and others are used regularly by radio broadcasters. Almost all live sports broadcasts on radio are backhauled to their main studios via ISDN connections.

NEXT-GENERATION NETWORK

The next-generation network (NGN) is a body of key architectural changes in telecommunication core and access networks. The general idea behind the NGN is that one network transports all information and services (voice, data, and all sorts of media such as video) by encapsulating these into IP packets, similar to those used on the Internet. NGNs are commonly built around the Internet Protocol, and therefore the term all IP is also sometimes used to describe the transformation of formerly telephone-centric networks toward NGN.

NGN is a different concept from Future Internet, which is more focused on the evolution of Internet in terms of the variety and interactions of services offered.

NGN Seminar in Fusion Technology Center by NICT (Japan) researcher.

A next-generation network (NGN) is a packet-based network which can provide services including Telecommunication Services and is able to make use of multiple broadband, quality of Service-enabled transport technologies and in which service-related functions are independent from underlying transport-related technologies. It offers unrestricted access by users to different service providers. It supports generalized mobility which will allow consistent and ubiquitous provision of services to users.

From a practical perspective, NGN involves three main architectural changes that need to be looked at separately:

- In the core network, NGN implies a consolidation of several (dedicated or overlay) transport networks each historically built for a different service into one core transport network (often based on IP and Ethernet). It implies amongst others the migration of voice from a circuit-switched architecture (PSTN) to VoIP, and also migration of legacy services such as X.25, frame relay (either commercial migration of the customer to a new service like IP VPN, or technical emigration by emulation of the "legacy service" on the NGN).

- In the wired access network, NGN implies the migration from the dual system of legacy voice next to xDSL setup in local exchanges to a converged setup in which the DSLAMs integrate *voice ports* or VoIP, making it possible to remove the voice switching infrastructure from the exchange.
- In the cable access network, NGN convergence implies migration of constant bit rate voice to CableLabs PacketCable standards that provide VoIP and SIP services. Both services ride over DOCSIS as the cable data layer standard.

In an NGN, there is a more defined separation between the transport (connectivity) portion of the network and the services that run on top of that transport. This means that whenever a provider wants to enable a new service, they can do so by defining it directly at the service layer without considering the transport layer – i.e. services are independent of transport details. Increasingly applications, including voice, tend to be independent of the access network (de-layering of network and applications) and will reside more on end-user devices (phone, PC, set-top box).

Underlying Technology Components

Next-generation networks are based on Internet technologies including Internet Protocol (IP)and multiprotocol label switching (MPLS). At the application level, Session Initiation Protocol (SIP) seems to be taking over from ITU-T H.323.

Initially H.323 was the most popular protocol, though its popularity decreased in the "local loop" due to its original poor traversal of network address translation (NAT) and firewalls. For this reason as domestic VoIP services have been developed, SIP has been more widely adopted. However, in voice networks where everything is under the control of the network operator or telco, many of the largest carriers use H.323 as the protocol of choice in their core backbones. With the most recent changes introduced for H.323, it is now possible for H.323 devices to easily and consistently traverse NAT and firewall devices, opening up the possibility that H.323 may again be looked upon more favorably in cases where such devices encumbered its use previously. Nonetheless, most of the telcos are extensively researching and supporting IP Multimedia Subsystem (IMS), which gives SIP a major chance of being the most widely adopted protocol.

For voice applications one of the most important devices in NGN is a Softswitch – aprogrammable device that controls Voice over IP (VoIP) calls. It enables correct integration of different protocols within NGN. The most important function of the Softswitch is creating the interface to the existing telephone network, PSTN, through Signalling Gateways and Media Gateways. However, the Softswitch as a term may be defined differently by the different equipment manufacturers and have somewhat different functions.

One may quite often find the term Gatekeeper in NGN literature. This was originally a VoIP device, which converted (using gateways) voice and data from their analog or digital switched-circuit form (PSTN, SS7) to the packet-based one (IP). It controlled one or more gateways. As soon as this kind of device started using the Media Gateway Control Protocol, the name was changed to Media Gateway Controller (MGC).

A Call Agent is a general name for devices/systems controlling calls.

The IP Multimedia Subsystem (IMS) is a standardised NGN architecture for an Internet media-services capability defined by the European Telecommunications Standards Institute (ETSI) and the 3rd Generation Partnership Project (3GPP).

Implementations

In the UK another popular acronym was introduced by BT (British Telecom) as 21CN (21st Century Networks, sometimes mistakenly quoted as C21N) — this is another loose term for NGN and denotes BT's initiative to deploy and operate NGN switches and networks in the period 2006–2008 (the aim being by 2008 BT to have only all-IP switches in their network). The concept was abandoned, however, in favor of maintaining current-generation equipment.

The first company in the UK to roll out a NGN was THUS plc which started deployment back in 1999. THUS' NGN contains 10,600 km of fibre optic cable with more than 190 points of presence throughout the UK. The core optical network uses dense wavelength-division multiplexing (DWDM) technology to provide scalability to many hundreds of gigabits per second of bandwidth, in line with growth demand. On top of this, the THUS backbone network uses MPLS technology to deliver the highest possible performance. IP/MPLS-based services carry voice, video and data traffic across a converged infrastructure, potentially allowing organisations to enjoy lower infrastructure costs, as well as added flexibility and functionality. Traffic can be prioritised with Classes of Service, coupled with Service Level Agreements (SLAs) that underpin quality of service performance guarantees. The THUS NGN accommodates seven Classes of Service, four of which are currently offered on MPLS IP VPN.

In the Netherlands, KPN is developing an NGN in a network transformation program called all-IP. Next Generation Networks also extends into the messaging domain and in Ireland, Openmind Networks has designed, built and deployed Traffic Control to handle the demands and requirements of all IP networks.

In Bulgaria, BTC (Bulgarian Telecommunications Company) has implemented the NGN as underlying network of its telco services on a large-scale project in 2004. The inherent flexibility and scalability of the new core network approach resulted in an unprecedented rise of classical services deployment as POTS/ISDN, Centrex, ADSL, VPN, as well as implementation of higher bandwidths for the Metro and Long-distance Ethernet/VPN services, cross-national transits and WebTV/IPTV application.

In February 2014 Deutsche Telekom revealed that its subsidiary Makedonski Telekom had become the first European incumbent to convert its PSTN infrastructure to an all IP network. It took just over two years for all 290,000 fixed lines to be migrated onto the new platform. The capital investment worth 14 million euros makes Macedonia the first country in the South-East Europe whose network will be fully based on Internet protocol.

In Canada, startup Wind Mobile owned by Globalive is deploying an all-ip wireless backbone for its mobile phone service.

In mid 2005, China Telecom announced its commercial roll-out of China Telecom's Next Generation Carrying Network, or CN2, using Internet Protocol Next-Generation Network (IP NGN) architecture. It's IPv6-capable backbone network leverages softswitches (the control layer) and protocols like DiffServ and MPLS, which boosts performance of its bearer layer. The MPLS-optimized architecture also enables Frame Relay and ATM traffic to be transported over a Layer 2 VPN, which supports both legacy traffic and new IP services over a single IP/MPLS network.

RADIO ACCESS NETWORK

A radio access network (RAN) is the part of a telecommunications system that connects individual devices to other parts of a network through radio connections. A RAN resides between user equipment, such as a mobile phone, a computer or any remotely controlled machine, and provides the connection with its core network. The RAN is a major component of wireless telecommunications and has evolved through the generations of mobile networking leading up to 5G.

An RAN provides access and coordinates the management of resources across the radio sites. A handset or other device is wirelessly connected to a backbone, or core network, and the RAN sends its signal to various wireless end points, so it can travel with other networks' traffic. A single handset/ phone could be connected at the same time to multiple RANs, sometimes called dual-mode handsets.

RAN components incude a base station and antennas that cover a specific region depending on their capacity. Silicon chips in both the core network as well as the user equipment provide RAN functionality.

The RAN controller is in control of the nodes connected to it. The network controller – which performs radio resource management, mobility management and data encryption – connects to the circuit-switched core network and the packet-switched core network, depending on the type of RAN.

The most recent evolution of RAN architecture divides the user plane from the control plane into separate elements. User data messages can then be exchanged by the RAN controller through one software-defined networking (SDN) switch and a second set through a control-based interface. This separation allows the RAN to be more flexible, accommodating for network functions virtualization (NFV) techniques such as network slicing and high MIMO that are necessary for 5G.

Radio Access Network Types

Types of RAN Include:

- GRAN, or GSM radio access network.
- GERAN, similar to the GRAN but specifies the inclusion of EDGE packet radio services.
- UTRAN, or UMTS (Universal Mobile Telephone System) RAN.
- E-UTRAN, or the Evolved Universal Terrestrial RAN.

References

- Kushnick, Bruce (7 January 2013). "What Are the Public Switched Telephone Networks, 'PSTN' and Why You Should Care?". Huffington Post Blog. Retrieved 11 April 2014
- Telecommunications-network, technology: britannica.com, Retrieved 18 February, 2019
- Weinstein, David (2004). The Forgotten Network: dumont and the Birth of American Television Temple University Press: Philadelphia, p. 16-17. ISBN 1-59213-499-8
- ARPANET, topic: britannica.com, Retrieved 19 March, 2019
- "Inductee Details – Donald Watts Davies". National Inventors Hall of Fame. Archived from the original on 6 September 2017. Retrieved 6 September 2017
- Radio-access-network-RAN, definition: searchnetworking.techtarget.com, Retrieved 20 April, 2019
- Jindal, R. P. (2009). "From millibits to terabits per second and beyond - Over 60 years of innovation". 2009 2nd International Workshop on Electron Devices and Semiconductor Technology: 1–6. doi:10.1109/EDST.2009.5166093. ISBN 978-1-4244-3831-0

Permissions

All chapters in this book are published with permission under the Creative Commons Attribution Share Alike License or equivalent. Every chapter published in this book has been scrutinized by our experts. Their significance has been extensively debated. The topics covered herein carry significant information for a comprehensive understanding. They may even be implemented as practical applications or may be referred to as a beginning point for further studies.

We would like to thank the editorial team for lending their expertise to make the book truly unique. They have played a crucial role in the development of this book. Without their invaluable contributions this book wouldn't have been possible. They have made vital efforts to compile up to date information on the varied aspects of this subject to make this book a valuable addition to the collection of many professionals and students.

This book was conceptualized with the vision of imparting up-to-date and integrated information in this field. To ensure the same, a matchless editorial board was set up. Every individual on the board went through rigorous rounds of assessment to prove their worth. After which they invested a large part of their time researching and compiling the most relevant data for our readers.

The editorial board has been involved in producing this book since its inception. They have spent rigorous hours researching and exploring the diverse topics which have resulted in the successful publishing of this book. They have passed on their knowledge of decades through this book. To expedite this challenging task, the publisher supported the team at every step. A small team of assistant editors was also appointed to further simplify the editing procedure and attain best results for the readers.

Apart from the editorial board, the designing team has also invested a significant amount of their time in understanding the subject and creating the most relevant covers. They scrutinized every image to scout for the most suitable representation of the subject and create an appropriate cover for the book.

The publishing team has been an ardent support to the editorial, designing and production team. Their endless efforts to recruit the best for this project, has resulted in the accomplishment of this book. They are a veteran in the field of academics and their pool of knowledge is as vast as their experience in printing. Their expertise and guidance has proved useful at every step. Their uncompromising quality standards have made this book an exceptional effort. Their encouragement from time to time has been an inspiration for everyone.

The publisher and the editorial board hope that this book will prove to be a valuable piece of knowledge for students, practitioners and scholars across the globe.

Index

CPSIA information can be obtained
at www.ICGtesting.com
Printed in the USA
LVHW060202060222
710369LV00002B/115

9 781632 388650